Life Evolving

Life Evolving

*Molecules, Mind,
and Meaning*

Christian de Duve

OXFORD

UNIVERSITY PRESS

2002

OXFORD
UNIVERSITY PRESS

Oxford   New York
Auckland   Bangkok   Buenos Aires   Cape Town   Chennai
Dar es Salaam   Delhi   Hong Kong   Istanbul   Karachi   Kolkata
Kuala Lumpur   Madrid   Melbourne   Mexico City   Mumbai   Nairobi
São Paulo   Shanghai   Singapore   Taipei   Tokyo   Toronto

and an associated company in Berlin

Copyright © 2002 by The Christian René de Duve Trust

Published by Oxford University Press, Inc.
198 Madison Avenue, New York, New York 10016

www.oup.com

Library of Congress Cataloging-in-Publication Data

De Duve, Christian.
Life evolving : molecules, mind, and meaning / Christian de Duve.
p. cm.
Includes bibliographical references.
ISBN 0-19-515605-6
1. Life—Origin.
2. Evolution (Biology)
3. Life (Biology)
I. Title.

QH325 .D415   20002
576. 8'8—dc21   2002075407

Permission for the reproduction of the images that appear in this book
was kindly granted by the artist, Ippy Patterson. The copyright
for these images rests with the artist.

9 8 7 6 5 4 3 2 1
Printed in the United States of America
on acid-free paper

CONTENTS

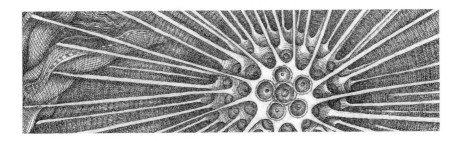

This preface is dedicated to all my colleagues,
past and present, at the Catholic University of Louvain

A WHIFF OF WOOD SMOKE

On a clear summer night almost 75 years ago, I was sitting, wrapped in a blanket, a scarf on my head, with a group of similarly clad youngsters circling a campfire. There was not a breath of wind. The flames rose straight toward an inky sky studded with stars. So did our voices joined in song, the only sound, with occasional crackling of burning wood, to break the silence of the night. All of a sudden, for a brief instant, light fused with darkness, song and silence became one, and I felt carried to another world, seized by intense emotion, suffused with a sense of un-fathomable mystery, feeling, beyond the infinite depths of space, the awe-some majesty of God.

Today, the boy scout of my reminiscence is an old man. What was then his future is now my past, a past that happened to coincide with the most dramatic burst of knowledge in the whole history of human-kind. The night sky of my youth has been explored to the outermost distance and earliest beginnings of the universe. The innumerable ap-pearances of matter have been reduced to a small set of elementary par-ticles and forces. Life itself has yielded its secrets. Its central mechanisms have been unravelled in intimate detail, and its history, which, as we now know, includes that of humankind, has been probed back to an origin lost in the mists of time.

As chance would have it, I did not live through those momentous events merely as a passive spectator. I was a privileged inside witness to them and even, to a modest extent, an active participant. This dizzying adventure was also a revealing discovery of reality, which totally upset the naïve set of beliefs from which had sprung the romantic mysticism of my childhood. Yet memory of that summer night never entirely faded away. It needs only a whiff of wood smoke to bring back the feeling of fervor and wonder that filled me at the time. The magic has gone, but not the sense of mystery.

EARLY INFLUENCES

I have recalled this childhood experience because it helps explain the tenor of this book, greatly influenced by my family background and early upbringing, especially in the religious domain. My family was Catholic, more by tradition and social conformism than by deep-felt conviction. We believed, without asking why, as a matter of course. Observance was faithful but largely perfunctory. We scrupulously refrained from eating meat on Fridays, attended mass every Sunday, confessed our sins regularly—or, in the case of the more tepid, at least once a year at Easter time, as prelude to the obligatory yearly Communion known as one's "Easter duty"—and we took care not to eat any solid food a minimum of twelve hours before receiving the sacred host at Communion. Religious holidays, such as Christmas or Easter, were duly celebrated. Church ceremonies underlined all major family events. We were baptized soon after birth and, later, when we had grown old enough to understand what was going on, confirmed by the local bishop, who took the trouble to come personally on that occasion. We married in church, which was also our last stop on the way to our final resting place.

Religion itself, however, hardly entered our home, except for the presence of a crucifix or other sacred image in every room. We never prayed together, read devout literature, or talked about religious topics. Politically, my father was mildly anticlerical and always voted liberal, not Catholic (the name of a political party at the time). I myself learned early, from spending holidays with German relatives who were Lutherans and with English friends of my parents who were nominally Anglicans but hardly bothered with religious practice, that one could reject the authority of the pope and skip mass on Sundays and still escape eternal damnation—that, at least, is what I believed—if one led a decent life. Rumor had it that some family members were actually unbelievers, perhaps even—perish the thought—Freemasons!

This broad-minded and tolerant family atmosphere did not prevent me from taking religion very seriously. In the Jesuit school I attended, Catholic doctrine was strictly imposed by highly intelligent and cultured Fathers, who described it as an unassailable, rational construction, firmly based on the teachings of Aristotle, as revised by Thomas Aquinas. Science, on the other hand, was poorly taught by teachers who distrusted it and took care not to present it as an opening to understanding the world. Only mathematics, thanks to its abstract character, escaped this neglect and was well expounded. Not knowing any better and not being inclined to question the wisdom of my teachers, I found this combination of reason and faith intellectually satisfactory and even appealing.

At the end of my "humanities," as classical high-school studies were called, there was never a moment of doubt in my mind or anyone else's that I should enter the Catholic University of Louvain, steering clear of its nearby rival the godless Free University of Brussels, founded in 1834 by a group of wealthy Freemasons with the aim of promoting, in direct opposition to dogmatic Louvain, a pernicious doctrine of "free thinking." In spite of my love for the classics, I opted for medical studies, mainly because I was attracted by the popular "man in white" image of the physician in the service of suffering humanity.

Through a combination of circumstances that have no place in this account, I discovered scientific research as a medical student and became so enamored with it that I abandoned clinical practice and specialized in biochemistry. At that same time, thanks to my first mentor, professor Joseph Bouckaert, to whom I remain deeply indebted, I discovered the scientific method of seeking truth, not by rational deduction from an a priori statement presented as incontrovertible, but by observation and experiment, continually questioned and subjected to the rigorous criterion of objective verification. It was an illumination that swept away, as by a tidal wave, the scholastic approach of the Jesuits and severely shook its doctrinal foundation. Claude Bernard[1] replaced Aristotle and Thomas Aquinas as my intellectual guide.

Notwithstanding this personal upheaval, at the end of my training I accepted an academic position at my alma mater. Given the almost complete inbreeding rampant in Belgian universities, there was no alternative if I was to stay in the country. I could have gone abroad—I had an attractive offer from the United States—but decided on Louvain in spite of my doctrinal qualms. There were several reasons for that decision, the main one being a sense of patriotic obligation, not yet seen as corny at the time, and the desire to help in the reconstruction of my war-ravaged country. I was no "rat leaving the sinking ship," was the way I put it to myself.

THE CATHOLIC UNIVERSITY

My qualms were not without justification. Probably not many of today's readers, even in Louvain, can readily imagine the kind of intellectual climate that existed at the Catholic University of Louvain in the time of my youth. In that venerable institution, founded in 1425 by Pope Martinus V, and within the conservative bourgeoisie from which much of its professorial staff was recruited, religious belief and practice were not so much an obligation as a deeply embedded way of life essentially taken for granted. The Belgian bishops made up the University's directing body, with as main prerogative the right to appoint professors (sons of professors and nephews of bishops were said to enjoy an undeniable advantage). The rector was a cleric, often of bishop rank. Students were carefully surrounded and watched, to the point that their lodging had to be approved by the vice-rector—a redoubtable individual, also a cleric—and a number of the town's more frivolous establishments were out-of-bounds for them. Female students enjoyed special protection, required to reside in colleges kept by nuns. All major events of academic life were celebrated in the main church, to which the professors marched in procession through the streets of the old city, garbed in elaborate gowns designed in the Middle Ages. Professors generally prefaced each course with a brief prayer. Much of the University's meager budget was fed from donations collected twice a year in churches throughout the country (with the reluctant collaboration of the local pastors, who did not gladly see a portion of their parishes' resources diverted for the benefit of an institution many of them viewed as subversive). Many professors returned their salaries to the University, relying for their living on private means or on revenues from lucrative practices as lawyers, physicians, or industry consultants made possible by their positions at the University.

Surprisingly, this almost-medieval framework hardly stifled academic freedom. Barring open opposition to the Church, it left open a wide field within which theologians could gleefully disagree with Rome and philosophers could defend widely dissenting views. Scientists were totally untrammeled in their teaching, supported by the theologians, who silenced censorship with the assertion that there could be no contradiction between truth and the Truth. Physicists taught the latest theories, including the Big Bang, actually discovered as the "primitive atom" by a clerical colleague, Monsignor Georges Lemaître. Biologists were free to explain living processes in terms of chemical reactions and bioenergetics in the framework of modern thermodynamics. They were not prevented from accepting biological evolution—not yet recognized by the Church at that time—and even its Darwinian explanation. Not all took advantage of this freedom, but it existed. In the medical faculty, neurologists did not hesitate to attribute to collective hysteria the allegedly miraculous apparitions of the Virgin Mary in the village of Beauraing (which did not prevent religious authorities from encouraging what they considered a commendable manifestation of popular devotion), whereas gynecologists, while obeying Catholic morality on abortion, for example, felt no scruple prescribing the "pill." Religious faith, however, was not a topic for open discussion, whatever doubts a person might privately entertain. I kept to this rule, except with a few intimate friends.

For the better part of my academic life, I had little difficulty avoiding controversial issues. There was no such thing as Catholic biochemistry; I had no party line to toe in my teaching, even less so in my research. After 1962, when I received a second appointment at the Rockefeller University in New York, my duties in Louvain became more episodic, largely restricted to research. Only in 1974 did I have to avoid a crisis when I was to be hailed as a Nobelist in science who was also a faithful son of the Church. On the whole, science kept me fully occupied, leaving little time for wider issues, even though these always remained at the back of my mind, kept in reserve for some later day, which, however, appeared too far in the offing to be of immediate concern.

THE MOMENT OF TRUTH

Time has caught up with me. Unlike many of my aging colleagues, I have given up laboratory research, ceased to advise investigators, and even stopped following the details of my discipline, devoting increasing amounts of my time and effort to more general problems, especially the origin and evolution of life. From these new preoccupations to "philosophical" considerations only one more step was needed, encouraged by the "whiff of wood smoke" whose memory has remained with me all my life. Slowly and cautiously, in the course of my reflections and writings, I moved closer to the ultimate question, managing, however, to skirt the final issue—the G word—in double-edged sentences that could be interpreted according to the reader's fancy. Between Teilhard and Monod, I ended up siding with Monod, but opting in favor of a meaningful world.[2] Between Pascal and Voltaire, I left the choice to the reader.[3]

No longer am I allowed to sit on the fence, even for the laudable intention of not hurting or shocking. This book is likely the last one I shall write. With apologies for such grandiloquent phrasing, it is to be my testament. I owe it to myself to express my true thoughts in it, with as much clarity and honesty as I can muster, whatever distressful surprise or disappointment such declarations may cause to a number of people. To all of those, I offer my apologies for what they may view as betrayal of my university, my colleagues, my friends, and my social milieu. At the same time, I beg them to read my testimony with attention and empathy. In actual fact, those who reproach me may perhaps not be as numerous as I fear. Today's Louvain is not the Louvain of fifty years ago. To give just one example, a Louvain physicist has recently, without encurring rebuke, as far as I know, made a public defense of atheism that goes much further than my vision of what, in the book, I call "ultimate reality."[4] This would have been unthinkable fifty years ago.

I suspect that many intellectuals who call themselves Catholics share my uneasiness but feel that religion is so necessary and beneficial that it is preferable not to rock the boat. As I mention at the end of this book, I have long felt the same and refrained from speaking out. But, as I approach the end of my journey, I have reached the conclusion that such an attitude is not defensible. Respect for truth takes precedence over the regard one may have for the opinions of others.

But most of my readers presumably will have no connection with Louvain University, Belgium, or the Catholic Church. I owe them an apology for this autobiographical account filled with details of interest only to myself. I hope they will understand, when reading the book, why I chose to bother them with seemingly trivial reminiscences and anecdotes. Perhaps my tale may strike a responsive chord in American readers. Whereas the world I depict has virtually disappeared from the European scene, the United States still harbors many fundamentalist institutions of so-called higher learning, in comparison with which even the Louvain of my youth would appear munificently liberal.

ABOUT THIS BOOK

The topic of this book is the history of life, from its earliest beginnings to the panoply of microbes, fungi, plants, and animals, including human beings, that envelops Earth today in a colorful web of throbbing life. As such, the book covers basically the same ground as my earlier *Vital Dust.*[5] But there are significant differences. First, there are additional chapters, devoted, for example, to biotechnologies and to extraterrestrial life and intelligence. Also, the language of the present book is less technical, more accessible than that of its predecessor. Most importantly, this book is unambiguously focused on explanation and on meaning. Rather than trying to describe even-handedly the present state of knowledge and to expose existing uncertainties or conflicts with the impartiality of an un-committed onlooker, I do not hesitate to argue matters and take sides. Especially, I tackle for the first time—or, at least, more explicitly than before—a number of sensitive questions, such as the role of chance in evolution, "intelligent design," religious beliefs, and the nature and in-tervention of God.

In writing this book, I have stretched myself far beyond the boundaries of my own competence. Even the greatest polymath, which I definitely am not, could not knowledgeably cover such vast ground. But the attempt deserves to be made and can't be just left to philosophers, historians, science writers, or other "generalists," who have no personal experience in science. Nor can the attempt be left, among those who have such experience, solely to physicists and cosmologists, who most frequently tend to venture into more general considerations but are often poorly acquainted with the life sciences. For better or for worse, I have taken the risk, apologizing to the experts for the many instances in which I presumed to trespass on their preserves and even had the temerity to offer my own opinion or interpretation on controversial questions.

This book is not a scholarly work in which every sentence is bolstered by appropriate references. Citations are mostly restricted to findings or statements of special interest, as reported in recent books or in nonspe-cialized journals, such as *Science, Nature,* or *Scientific American,* my main sources of information in the last few years. More complete coverage of the literature may be found in my earlier *Vital Dust* (1995) and *Blueprint for a Cell* (1991).[6] In addition, many details of more specialized nature have been left out of the main text for easier readability and are given in separate notes grouped together at the back of the book. Readers with an appetite for more solid fare are referred to these notes, which can to some extent be read as appendices.

Before closing this preface, I owe the readers one more explanation. I have written this book more or less in parallel in my two mother languages: French, in which I was educated and most often converse; and English, in which, being born in England, I was immersed from my earliest childhood and which has become, for all practical purposes, my main scientific means of expression and even thinking. A book written in this manner is a strange chimera, not only stylistically—which can be corrected with appropriate assistance—but also conceptually. One thinks differently in French and in English. Readers belonging to one culture or the other may find this somewhat disconcerting. Unfortunately, that's the way the author's brain was wired.

ACKNOWLEDGMENTS

I owe a special debt of gratitude to my old friend Neil Patterson, who has not only applied all his editorial skills to pulling together the text into a reasonably readable whole, trimming it of most of its Gallicisms—I have insisted on keeping some to preserve my identity—and of many flowery, pompous, incautious, obscure, or misleading phrases. He has also addressed the book's actual substance and helped me greatly in keeping the science on a rigorous course and in clarifying my own thoughts and ideas. In addition, Neil has shared my writings with his wife, Ippy, a wonderful artist who already contributed the trees to my *Vital Dust*[7] and has now given the present book a unique flavor with her beautifully delicate drawings. I am deeply grateful to her.

A number of colleagues have kindly read parts of the book lying in their own area of expertise and given me the valuable benefit of their comments and criticisms—which I confess I have chosen to follow as I saw fit. I am particularly grateful, in this connection, to Jacques Berthet, Susan Blackmore, Ivar Giaever, Henri-Géry Hers, Miklos Müller, Sue Savage-Rumbaugh, Jill Tarter, and Marc van Montagu.

I also acknowledge with grateful thanks the valuable assistance of my New York secretary, Anna Polowetzky (Karrie), and her son Michael, who have helped me with many phases of the book, including biographical searches. In Brussels, Monique Van de Maele has also provided much useful aid. I have been particularly fortunate to have this book published, under their customary high standards, by Oxford University Press. I am pleased to express my appreciation to Kirk Jensen and his colleagues for their competent and dedicated professionalism. The editorial assistance of Catherine Humphries has been especially valuable.

As in all my past endeavors, this one could not have been brought to successful conclusion without the devoted support and forbearance of my dear wife Janine. All I can do in return is to use this preface to acknowledge publicly and with loving gratitude her invaluable contribution.

Néthen and New York, Spring 2002

Life Evolving

Look at the five "words" below, knowing that they were written with an alphabet of 20 letters:

ILDIGDASAQELAEILKNAKTILWNGP
GLDIGPDSVKTFNDALDTTQTIIWNGP
GLDVGPKTRELFAAPIARAKLIVWNGP
GLDCGTESSKKYAEAVARAKQIVWNGP
GLDCGPESSKKYAEAVTRAKQIVWNGP

If I were to tell you the words were typed separately by five different monkeys, would you believe me? Not if you have taken more than a passing glance at them. "All five words end with WNGP," you would point out to me, "and for monkeys hitting keyboards independently, this cannot be." Actually it can. But the probability of such a coincidence is one in 655 billion billions. You would need a pretty large number of monkeys for five of them to have a reasonable chance of coming up with the same word ending. Surely, a more likely possibility is that the monkeys cheated. They *copied*!

Actually, the fraud is even more flagrant than appears at first sight. If you look more closely, you will see that four other letters, in addition to the terminal four, are the same in all five words (LD in position 2 and 3, G in position 5, and I in position 22). This lowers the odds of a fortuitous coincidence to one in 429,500 billion billion billion billions. Trillions of planets like ours could not possibly provide enough monkeys. And this is not all. Five other letters are the same in four out of the five words (G in position 1, S in position 8, A in position 13, and AK in positions 19–20). Even more striking, the two last words have 25 out of 27 letters in common; they differ only in positions 6 and 17. There can be no doubt. If monkeys there were, they most certainly did not hit their typewriters' keys at random.

The words shown are not inventions. They represent real things, fragments of molecules called proteins, which are very long chains of up to several hundred units called amino acids, of which 20 different kinds are used in the assembly of the chains. Each word represents the sequence of a 27-amino acid piece (each letter standing for a given kind of amino acid) present somewhere in the heart of a large protein molecule containing more than 400 amino acids. This protein is an enzyme, or biological catalyst, known as phosphoglycerate kinase, PGK for short. PGK is a key participant in one of the most fundamental processes that take place in living organisms, the conversion of sugar to alcohol (or lactic acid), which occurs in virtually all forms of life, whether microbes of various sorts, plants, molds, or animals (including humans).

Now comes the central piece of information, which explains why the words serve as an introduction to this book. The five structures shown belong to the PGKs of five widely different organisms. The first one belongs to *Escherichia coli*, or colibacillus, a common microbe that we all harbor in our gut. The others are from the wheat, fruitfly, horse, and human PGKs, respectively:

Colibacillus:	ILDIGDASAQELAEILKNAKTILWNGP
Wheat:	GLDIGPDSVKTFNDALDTTQTIIWNGP
Fruitfly:	GLDVGPKTRELFAAPIARAKLIVWNGP
Horse:	GLDCGTESSKKYAEAVARAKQIVWNGP
Human:	GLDCGPESSKKYAEAVTRAKQIVWNGP

What our monkey parable has brought to light is that the similarities among the PGKs of our sample organisms could not possibly be due to chance. A possibility could be—this, no doubt, would be the "creationist" view—that the similarities betray the intervention of a "hidden hand." But, in that case, why the differences? Why, for example, does the human sequence differ from the fruitfly sequence in twelve amino acids and from the horse sequence in only two? No, the explanation given above for the monkeys is the correct one. The sequences show similarities because they were *copied*. And, they show differences because occasional copying mistakes were made. Thus, two mistakes would have been made in the horse and human lineages, twelve in the human (or horse) and fruitfly lineages, since their respective PGKs started being copied separately. Or, as shown graphically:

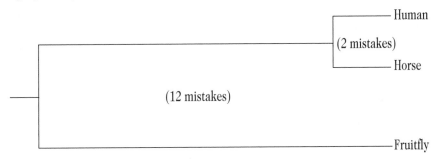

Make the additional assumption that it took some 40 million years, on an average, for one mistake to be made, and you get the following:

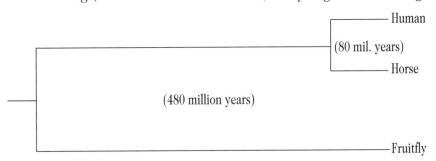

This, very roughly, is what paleontologists have long been telling us on the strength of fossil evidence. Humans and horses are derived from a common mammalian ancestor from which they diverged some 80 million years ago. The mammals themselves and the insects (the parent group of fruitflies) separated from a common ancestral form roughly 500 million years ago. What is new is that we can now estimate evolutionary times in terms of copying accidents (mutations) and that we can extend such estimates to lineages that have left no fossil remains. Also, we know how the copying takes place. It does not involve the protein molecules themselves, as suggested for simplicity's sake; it involves the DNA genes that encode the amino acid sequences of the protein molecules. For the purpose of our argument, it amounts to the same thing.

More will be said about this fascinating topic in Chapter 7. The main point, for the time being, and the reason for this Introduction, is that there is now overwhelming evidence that *all known living beings are descendants through evolution from a single ancestral form of life*. Many cogent reasons support this affirmation. Its most convincing proof is provided by the molecular sequencing results.[1] Even the very limited data presented in this Introduction should suffice to demonstrate the kinship among the five organisms mentioned (which, it should be noted, include us and the colibacilli of our intestinal tract). All the other available data—and their number is ever increasing—have confirmed this kinship and extended it to every other organism so far investigated. This fact is now so well established that researchers would be overjoyed if even one exception could be found—whether on Earth or elsewhere—because it would point to a second, independent origin for life.

1. What Is Life?

U NTIL RECENTLY, the answer to the question "What is life?" posed no problem. Life, it was said, is "animated matter," from the Latin *anima*, soul. This, of course, was no explanation at all. It simply attributed to the soul, or vital spirit, all that was not understood about life. Nevertheless, vitalism, as this doctrine is called, maintained a foothold until well into the twentieth century, often—quite misguidedly—in connection with religious beliefs. This was especially true in France and other French-speaking countries.

The great Louis Pasteur was a confirmed vitalist. So was the philosopher Henri Bergson, winner of the 1927 Nobel prize in literature and author of *L'évolution créatrice*, in which evolution is depicted as propelled by an "*élan vital*," a vital surge. It was the case also for the physicist Pierre Lecomte du Nouy, who coined the term "*téléfinalisme*" to designate what he perceived as the innate ability of living organisms to act purposefully, in opposition to the second law of thermodynamics. When I was a student at the Catholic University of Louvain, in Belgium, all my biology teachers were vitalists, with the exception of the physiology professor, Joseph Bouckaert, in whose laboratory I had the good fortune of taking my first steps in science and who was a staunch mechanist. Which did not prevent him from attending church every Sunday and behaving like a perfectly good Christian. In his view, science and religion belonged to different domains, which, exhibiting no overlap, could not contradict each other.

Today, vitalism has few adherents,[1] as more and more of the remarkable properties of living organisms are being explained in terms of physics and chemistry. In turn, attempts at defining life call increasingly on these disciplines. In 1944, the Austrian physicist Erwin Schrödinger, world-renowned for the development of wave mechanics, addressed the question in a booklet titled *What Is Life?*, which was highly influential at the time. He perceptively singled out two properties as particularly characteristic of living beings: 1) their ability to create order out of disorder by exploiting external energy sources, feeding on what he called "negative entropy"; and 2) their capacity to transmit their specific blueprint from generation to generation, which property Schrödinger, who knew nothing of DNA, attributed to an "aperiodic crystal."

More recently, evolutionists, such as Britain's Richard Dawkins, have highlighted the paradigm of the "selfish gene," a powerful image intended to illustrate the notion that the genes are the ultimate targets of natural selection. Theorists, like Stuart Kauffman, long associated with the famous Santa Fe Institute, where so-called artificial life is being created by computers, insist on "self-organization" as a central property of life. My Belgian colleague Ilya Prigogine sees life as an example of those "dissipative structures" of which he has made a detailed theoretical study. Thus each, depending on personal interests, biases, and training, has his or her answer to the question "What is life?" Mine is simple.

LIFE IS WHAT IS COMMON TO ALL LIVING BEINGS

This answer is not a tautology, as it allows many attributes to be excluded from the definition of life. There is no need for green leaves, or wings, or arms and legs, or a brain, to be alive. It is not even necessary to be made of many cells. Hosts of living organisms consist of single cells. The simplest among them, namely the bacteria, lack a central nucleus and most of the other structures that can be seen inside the cells of more evolved organisms; so those cell features are also not requisites of life. What remains is what we humans have in common with the colibacilli in our gut. It is still a lot.

We and colibacilli, together with all other living beings, are made of cells, which are constructed with the same substances. We build our constituents by the same mechanisms. We depend on the same processes to extract energy from the environment and convert it into useful work. Most telling of all, we use the same genetic language, obey the same code. There are differences, of course. Otherwise, we would all be identical. But the basic blueprint is the same. There is only one life. In fact, all known living beings descend from a single ancestral form. We and colibacilli are distant cousins; very distant, but indubitably related.

A mere 50 years ago, these notions were still very dim and backed by little evidence. Today, it is no exaggeration to state that we know the secret of life. In just half a century, humankind has made the biggest leap in knowledge in its whole history. This revelation has come to us from the advances of cell biology, biochemistry, and molecular biology.

Living Beings Are Made of Cells

A literary person knowing nothing of biology might find this affirmation puzzling. Indeed, the word "cell" stands for a small room, a cubicle. One speaks of the cell of a monk or a prisoner. But what has life to do with such chambers? The explanation is to be found in a richly illustrated book, *Micrographia*, published in 1665 by the English physicist Robert Hooke, who built one of the first microscopes. Among the images reproduced in this work, there is a drawing of a thin slice of cork, in which Hooke distinguishes a fine, honeycomb structure, consisting, he says, of "microscopic pores," or "cells." The term has remained, to designate, not the cavities of cork, but the small bodies that occupy them in the living bark; it was adopted in 1837 by two German scientists, the physiologist Theodor Schwann and the botanist Matthias Schleiden, who proposed what is known as the generalized cell theory, according to which all living beings consist of cells.

Another German biologist, Rudolf Virchow, drove the generalization one step further in his classic opus *Die Cellularpathologie* (1855), in which he writes: "*Omnis cellula e cellula*" (all cells arise from cells). This was a paraphrase of the aphorism by the celebrated seventeenth-century English physician William Harvey, discoverer of blood circulation and explorer of animal generation: "*Ex ovo omnia*" (all [living beings] come from an egg).

Virchow's rule, as it is now known, suffers no exception. Everything that lives is made of one or more cells. And all cells come from other cells, by growth followed by division. This applies to our cells. The cells of our skin, our liver, our brain, and all our other organs originate, by successive divisions, from a single cell, the fertilized egg. The outcome, however, is not a simple collection of identical cells, a clone of the egg cell; it is a true organism. During embryological development, the cells, as they undergo successive divisions, become progressively differentiated and organized into tissues and organs. The egg cell itself arose from the fusion of two cells, the maternal and paternal germ cells, which were themselves decendants of egg cells. One can thus go back, by uninterrupted continuity, from any of our cells to the very first cells that existed on Earth. What is true for us is true also for any other living being. From cell to cell, all forms of life are descendants of those first cells.

How did the first cells originate from materials that were not organized as cells? How, in turn, did they give rise to the whole panoply of living beings that populate Earth today? Those are the two central questions raised by the history of life. To them must be added one more problem, of truly mind-boggling complexity: how does a fertilized egg cell produce, by multiplying, a harmoniously developed organism that closely resembles the donors of the germ cells from which that egg cell arose? In the chapters that follow, I shall try to sketch out the still-fragmentary answers science has, largely in the last 50 years, provided to these fundamental questions, which raise the existential problem of our presence here on Earth.

Cells are microscopic globules, of dimensions measured in thousandths of a millimeter. The human body contains trillions (thousands of billions) of cells. Of gelatinous consistency, cells have shapes that vary according to the internal and external constraints to which they are subjected. In the absence of such constraints, they tend to adopt a spherical shape.

To make a cell, there is first needed an envelope that serves as a border and, like all borders, isolates what it surrounds, while providing controlled entry and exit ports for the indispensable exchanges of matter that cells maintain with the outside. These functions are carried out by the *cell membrane*, a tenuous pellicle hardly ten-millionths of a millimeter thick. As fragile as a soap bubble, which it resembles in some of its physical properties, the cell membrane is most often, but not invariably, protected and bolstered by an external, rigid *wall*. In pluricellular organisms, this wall is replaced by scaffoldings, sometimes very elaborate, that shore up the tissues and delineate their architecture. Hooke's original cells were nothing but the putrefaction-resistant remnants of such scaffoldings.

The inside of the cell is occupied by a kind of semiliquid gel, called *cytoplasm*, showing little structure in certain cells and filled, in others, with a variety of granules, vesicles, and other entities, which constitute distinct organelles (small organs) and functional systems. The chemical processes that support the maintenance and growth of the cell take place largely in the cytoplasm.

Finally, the third indispensable component of any cell is what might be called its information center, the site in which are stored the instructions that command all that goes on in the cell and from which these instructions are issued. This information is written, in a coded chemical language, in one or more circular or rod-shaped structures, called chromosomes. These are naked and in direct contact with the cytoplasm in the simplest cells, those of bacteria. In the cells of more complex living beings, the chromosomes are confined within a central enclosure separate from the cytoplasm, the *nucleus*. On the basis of this distinction, the cells of bacteria are designated *prokaryotic* (from the Greek *karyon*, kernel), and the others *eukaryotic* (from the Greek *eu*, good).

First known by the diseases caused by some of them, bacteria are ubiquitous. Their sizes are small, on the order of one-thousandth of a millimeter, and their internal organization is rudimentary. They are found in a wide variety of environments. Most have no pathogenic effect; many are useful. It is estimated that bacteria represent, collectively, a mass equivalent to that of all trees and other plants combined.

Other unicellular microbes exist, but of eukaryotic nature, much larger and more complex than bacteria. Formerly subdivided into protozoa (primitive animals) and protophytes (primitive plants), these organisms are now grouped under the name of protists. They include the largest— up to the point of being sometimes visible with the naked eye—and most elaborate known eukaryotic cells. Certain severe illnesses, such as malaria and sleeping sickness, are due to protists. Many other protists are harmless and proliferate in many sites. They teem in the waters of puddles and ponds, to the delight of all those who have contemplated through a magnifying glass these "animalcules," thus named by Antonie van Leeuwenhoek, a Dutch contemporary of Robert Hooke and inventor, like him, of one of the first microscopes.

All pluricellular organisms (that is, plants, fungi, and animals, including humans) are made of eukaryotic cells. Mostly some 20 to 30 thousandths of a millimeter in size, these cells are all constructed according to the same general blueprint, the biggest differences being those existing between plant cells and the others. Within a given organism, the basic blueprint is subject to a number of variations linked to functional specializations. The number of different cell types in an organism increases with rising complexity. It reaches some 220 in the higher mammals, whose developmental programs are particularly complicated.

Living Beings Are Chemical Factories

Making life is what life is all about. What is thus made bears a remarkable similarity to what exists, leading many of those who have reflected on the phenomenon to insist that the central characteristic of life is the ability to follow a blueprint. This is, no doubt, an all-important feature of living organisms, and it will be considered in the next chapter. But blueprints are useless without builders. And life is built by *chemical* mechanisms. There can be no attempt at understanding life without the language of chemistry. This is all the more true because even biological information depends on chemistry. Unfortunately, few of us are familiar with even the basic elements of chemistry, in spite of the leading role of chemical industries in our technological civilization. Every effort has been made in this book to avoid technical details. But for life to be made understandable in modern terms, a minimum of chemistry has to be included.

THE FACTORIES NEVER STOP

Growth and multiplication are the most evident manifestations of life's self-building property. This activity is exercised also in a steady state, where nothing seems to change; but, in reality, construction work of all sorts continually takes place, offsetting an equivalent degree of decay. Indeed, breaking down life is as much a central characteristic of life as is making it. The two activities are inseparable and, together, account for the turnover, or renewal, of biological constituents. A bare few weeks after their birth, cells that have not multiplied and seem entirely unchanged start resembling those old houses that have maintained the same shape in the course of centuries but have had many boards, bricks, tiles, and window panes replaced.

This remarkable phenomenon and its astonishing magnitude have been revealed by the use of radioactively labelled substances, that is, substances in which certain atoms have been replaced by their radioactive counterparts (isotopes), carbon of atomic mass 12 by the radioactive carbon of atomic mass 14, for example, or hydrogen of atomic mass 1 by the radioactive hydrogen of atomic mass 3 (tritium).[2] When an organism is briefly provided—pulsed is the technical term—with some foodstuff containing radioactive atoms, these are found (detected by their radioactivity) to be rapidly incorporated into some biological constituents, from which they subsequently disappear just as swiftly, replaced by non-radioactive atoms as soon as the labelled foodstuff ceases to be supplied. Thus, even though the total amount of the constituents present has remained constant during the time of the experiment, their dynamic state, continually subject to breakdown and synthesis, has been brought to light by the use of labelled foodstuffs.

What, now, about the mechanisms whereby living organisms exercise their remarkable self-building ability? As for all chemical syntheses, three conditions must be fulfilled: raw materials, energy, and, in almost all cases, catalysis. In addition, a chemical factory that constructs itself must be able to combine the products of its industry according to a definite plan.

RAW MATERIALS

Let us first consider the case most familiar to us, our own. We derive our raw materials from our food. This is made easy because the animals and plants on which we feed are constructed with the same building blocks as our own tissues. The building blocks themselves are typically molecules of small size, made of carbon, hydrogen, and, most frequently, oxygen. They often contain nitrogen, sometimes sulfur. The number of atoms per molecule rarely exceeds 30, which gives molecular masses generally lower than two hundred times the mass of the hydrogen atom. For a chemist, these are simple substances, easy to make in the laboratory. On the whole, little more than 50 different kinds of such simple substances—mostly sugars, amino acids, nitrogenous bases, fatty acids, and a few other more specialized compounds—account together for more than 99 percent of the organic matter of any living being. To this must be added water, always the principal component, and a certain number of mineral elements, including sodium, potassium, chlorine, calcium, magnesium, iron, copper, and a few others.

What makes the difference, say, between our foodstuffs and ourselves is the way building blocks are assembled—somewhat like pieces of a

Lego game—into large molecules (macromolecules), mostly consisting of long chains in which as many as several hundred pieces, if not thousands, are joined together end to end. These chains are often folded and twisted into three-dimensional assemblages whose characteristic shapes are critically important for the biological properties of the substances. Both the organisms from which we derive our food and our own tissues—as well as all other living beings—are made of substances built according to the same general models but differing in the sequences of the chains—that is, in the order in which various building blocks follow each other along the chains. For this reason, and also because large molecules do not readily enter the organism, we cannot use food macromolecules directly. We must first dismantle them into their constituent building blocks. This process, called *digestion*, takes place in the alimentary tract. Intestinal absorption then transfers the products of digestion into the bloodstream, which, in turn, conveys them to all the cells of the body.

There, in the cells, the small molecules derived by digestion from food macromolecules enter a kind of chemical whirlpool called *metabolism*, in which thousands of reactions allow the substances present to be modified in various ways. It is from this metabolic pool that our cellular factories draw the materials with which they manufacture the characteristic constituents of our cells and tissues. The Lego pieces are thus reassembled into new structures proper to our organism. If, as is usually the case, certain necessary pieces are inadequately provided, they are made from others by metabolism, which also furnishes the energy required for the assembly reactions and other forms of work.

Feeding, digestion, absorption, metabolism, assembly: those are the obligatory steps in the transmutations whereby, for example, a baby makes human tissues from cow's milk. The same five steps allow the cow to make milk from grass. But here the food chain reaches the end. Grass does not eat in the usual sense of the word. It makes grass from simple inorganic substances: water from the soil, carbon dioxide from the atmosphere, a source of nitrogen, most often nitrate, and a few mineral salts, with, in addition, the indispensable source of energy, which is sunlight.

The examples just described can be generalized. There are two classes of living beings: those, like babies and cows, that feed on other living beings, and those, like grass, that utilize nonliving sources. The former, known as *heterotrophs* (from the Greek *hêteros*, other, and *trophê*, nourishment), include all animals and fungi and many microbes, both protists and bacteria. All use their foodstuffs by the same mechanisms. Even unicellular heterotrophs do so. Protists depend on special feeding processes whereby food is internalized and digested in intracellular pockets known as lysosomes. Bacteria digest their food extracellularly and then absorb the digestion products.

The organisms that make their constituents from nonliving sources are designated *autotrophs* (from the Greek *autos*, self, and *trophê*, nourishment). Most are *photosynthetic*, that is, derive the energy they need from light (*phôs* in Greek). They comprise the pluricellular plants and unicellular algae, which are eukaryotes, and photosynthetic bacteria. The last two are the main constituents of phytoplankton, the vast, life-generating solar screen that floats on the surface of oceans and initiates the marine food chain. Some autotrophic bacteria, called *chemosynthetic*, do not need light; they obtain their energy from mineral chemical reactions, such as the conversion of sulfur to sulfate or the production of methane (CH_4) from carbon dioxide and hydrogen. This property allows chemosynthetic organisms to develop in unlikely ecological niches, such as abyssal hydrothermal vents or deeply buried rocks.

Autotrophs, as well as heterotrophs, when cut off from their energy supply (plants in the dark, fasting animals) are able to cover their needs for a certain amount of time by subsisting on their stores (of starch or fat, for example) and even part of their active substance. All are now familiar with those shocking images of fleshless bodies, veritable living skeletons, who managed to survive in the Nazi horror camps or, closer to us today, try to subsist in regions ravaged by famine or war, awaiting the arrival of life-saving food supplies.

ENERGY

Biological self-constructions require energy. So do the other kinds of work—mechanical, electrical, osmotic, etc.—carried out by living beings. In the last analysis, the main source of this energy is sunlight, which directly supports all green plants and other photosynthetic organisms and, by way of the alimentary chain, all the other organisms that ultimately depend on food supplied by the photosynthetic ones. The baby fed cow's milk, for example, derives its energy from sunlight by way of the grass eaten by the milk-providing cow. Only chemosynthetic bacteria (those autotrophs that derive their energy from mineral chemical reactions) and the organisms that feed on them do not depend on sunlight. At present, only a small part of the living world belongs to this category, which, however, may be of great significance for the origin of life.

The biological utilization of light will be more readily understood if we first look at how we and all other *aerobic* (living in air) heterotrophic organisms—that is, animals, fungi, and many protists and bacteria—meet our energy needs. The operative word is combustion; more technically, oxidation,[3] the energy-producing interaction of certain substances with oxygen. In this respect, we resemble motor cars, which run on the combustion of gasoline; or heat power plants, which manufacture electricity by burning coal, oil, or natural gas. The fuel, in our case, consists of components of the metabolic pool (derived themselves from foodstuffs). Here, however, the analogy ends. Vital combustions are cold; and the energy they release is not utilized in the form of heat, a phenomenon that would be impossible in living cells, where temperature differences are negligible. Instead, this energy serves to drive a central chemical generator that, in turn, powers most forms of biological work. The nature of this generator will be considered below.

In cellular combustions, as in those we are familiar with, oxygen is used to convert the carbon of organic substances into carbon dioxide (CO_2) and their hydrogen into water (H_2O). Exactly the opposite takes place in photosynthesis.[4] What green plants do with the help of light energy is simply to reverse oxidations. Starting from carbon dioxide and water, the plants manufacture a sugar of formula $(CH_2O)_6$, throwing off the excess oxygen (one molecule of O_2 for each molecule of CO_2 used) into the atmosphere. The fuel is thus regenerated at the expense of the products of its oxidation, the required energy being supplied by light. Everything else, or almost, takes place as in aerobic heterotrophy—which is the state plants change to in the dark, subsisting on their reserves. Many photosynthetic bacteria act like plants, but a few rely on more primitive reactions that lead to sugar synthesis without the release of oxygen. As to nonphotosynthetic (chemosynthetic) autotrophic bacteria, they accomplish the same kind of syntheses with energy provided by the oxidation or other transformations of mineral substances.

The two processes—the dominant form of photosynthesis, which consumes carbon dioxide and produces oxygen, and biological oxidations, which consume oxygen and produce carbon dioxide—tend to balance each other worldwide, so that the levels of the two gases in the oceans and atmosphere remain constant. In recent years, however, this balance is being threatened by the ever-increasing consumption of fossil fuels combined with the progressive shrinking of forested areas. For oxygen, which represents 21 percent of the atmosphere, the disturbance is negligible. But for carbon dioxide, which makes up little more than 0.03 percent of the atmosphere, the rise caused by increased human-caused production and lower photosynthetic consumption has already become significant. There is increasing evidence that this phenomenon is beginning to cause a warming of Earth, due to the greenhouse effect.[5] If the present trend is allowed to continue, it could lead to the flooding of large coastal areas through melting of polar ice and to other catastrophic consequences for the environment. Awareness of these risks has reached higher levels of government. But the required measures will be very difficult to take, especially in view of the growing opposition to nuclear energy, at present the cheapest and most readily available substitute for fossil fuel consumption.

Oxidations, though playing a preponderant role, are not the only energy-supplying reactions of heterotrophic organisms. Some organisms, called *anaerobic* (living without air), can power the central generator by means of chemical processes that do not involve oxygen, for example, the fermentation of sugar to alcohol or lactic acid.[6] Some of these organisms are facultatively anaerobic; they can develop in the presence or absence of oxygen. Such is the case with yeasts, which (under oxygen-free conditions) make most of the alcohol we consume. Even our own muscles can be transiently anaerobic. The cramps that sometimes affect athletes are due to the lactic acid produced anaerobically in their muscles when these are inadequately supplied with oxygen during strenuous effort. Some organisms, such as the bacillus of gaseous gangrene, are obligatorily anaerobic. They can develop only in the absence of oxygen and are killed by this substance. We shall see later that this fact is of crucial importance for the origin and evolution of life. (It also explains why gaseous gangrene can readily be prevented simply by incising wounds and exposing them amply to the oxygen in air.)

What about the central generator powered by energy-yielding oxidations and fermentations? It is a chemical machinery that produces a compound called adenosine triphosphate (ATP), the fruit of the union of adenosine diphosphate (ADP) and inorganic phosphate (P_i). Never mind the exact chemical nature of these substances.[7] What counts is that energy is required to combine ADP with phosphate; this energy is quantitatively returned when ATP is split back into ADP and phosphate. These two reactions serve universally in the transfer of energy from metabolism to biological work.

Let us consider work first. Most forms of biological work are powered by the splitting of ATP,[8] with the help of specialized machineries, or *transducers*. Our muscles, for example, and other biological motor systems are driven by ATP. So, most often, are the cell systems involved in the specific import or export of substances; the generators that produce electricity in torpedo fish, electric eels, and the nervous system of animals; the light organs of fireflies and glowworms; and, of course, all the many chemical processes involved in biosynthetic constructions.[9] For those reasons, ATP is sometimes referred to as the "fuel" of life. This expression could be misleading, however. ATP is not burned, but split, to provide energy.

As to the reactions whereby ATP is assembled with the help of metabolic energy, they depend on special *couplings* between certain metabolic reactions[10] that produce energy and the energy-consuming creation of a chemical bond between ADP and phosphate. The two processes are linked in such a way that the energy-producing process cannot take place without driving the other. A similar thing happens in our engines. There is coupling between the combustion of gasoline and the propulsion of a motor car or between the energy-yielding process in a power plant—be it fuel combustion, falling water, or nuclear fission—and the rotation of an electric generator. But those are mechanical couplings. Biological couplings are chemical.

ATP is the universal vehicle of energy in the living world. Its role is analogous to that of electricity in the economy. Electricity, produced in central power plants and transported by conductors, drives all sorts of machines and appliances that convert it, with the help of appropriate transducers, into mechanical work, heat, light, and, sometimes, physical or chemical work. With ATP, the vehicle is different—a substance circulating by diffusion instead of electric current—and the couplings and transducers are of a different nature. But the principle is similar.

CATALYSIS

Biosynthetic assemblies, metabolic transformations, and bioenergetic couplings involve a very large number of chemical reactions, of which virtually none would take place if the participating substances were merely mixed together. Living beings carry out these reactions thanks to the mediation of specific catalysts. This term, coined by the great Swedish chemist Jakob Berzelius, designates a substance that helps a chemical reaction to take place, without itself being consumed in the reaction. Biological catalysts are called *enzymes* (recalling the fact that they were first discovered as agents of fermentation in yeast, which is called *zymê* in Greek).

Enzymes do truly stupendous things! They selectively fish out, by means of what are known as *binding sites*, the substances on which they act—the technical term is substrates—from the metabolic pool, a highly complex mixture containing up to several thousand different substances, most of them at very low concentration. Each kind of enzyme selects its own particular substrate or substrates from the metabolic pool. Substances thus caught end up accurately positioned with regard to another special part of the enzyme molecule, called the *active center*, that brings about their modification. This may be the splitting of a substance into two pieces, or the joining of two pieces into a single entity, or, more frequently, an exchange of electrons or chemical groups between two substances. As soon as the reaction is completed, its product or products detach from the enzyme surface, leaving the sites open for a new round. Thousands of such cycles—the record exceeds half a million—may take place every second on the surface of a single enzyme molecule!

Hundreds, if not more, of such reactions, each involving a different kind of enzyme and different participating substances, take place side by side in even the most primitive of living cells. In most cases, the products of certain reactions serve as substrates for others, thus linking consecutive reactions into a variety of chains, which may be linear, branched, or cyclic. Called *metabolic pathways*, these chains of reactions mediate all the chemical modifications that take place in cells. The metabolic pool consists essentially of all the substrates and products of the enzymes present. A few substances feed into this pool from the outside; a few end products are discharged from it as waste.

Any living being is a reflection of its enzyme arsenal. We are and do what our enzymes permit. This is so true that the absence of a single enzyme—as a consequence of a genetic deficiency, for example—often suffices to completely disorganize metabolism, to the point of severely endangering survival. This is the explanation for many hereditary diseases, to which, in the early twentieth century, the British pediatrician Sir Archibald Garrod gave the imaginative name of metabolic errors. Similarly, poisons and drugs frequently owe their biological effects to their ability to block certain enzymatic reactions.

Faced with these facts, you begin to get an idea of the power of enzymes and their significance for life. You also wonder at the nature of chemical structures that can create such a wide spectrum of finely tuned configurations as make up all the various binding sites and active centers present in enzymes and that can, in addition, have these binding sites and active centers arranged with pinpoint accuracy in the relative positions needed for the chemical reactions to take place. Clearly, substances capable of that kind of jugglery must belong to a class of substances with exceptionally rich and versatile properties.

These substances are the *proteins*, which are, indeed, the most complex substances found in living beings. Like other natural macromolecules, proteins are long, very thin strings made by the linking end-to-end of a large number—most often several hundreds—of pieces. Remember, we saw a small fragment of an enzyme protein in the Introduction. What makes proteins particularly complex is that 20 different kinds of pieces serve in the making of the strings and that these pieces, which belong to the group called *amino acids*, show an extraordinary variety of physical-chemical properties. Some amino acids carry a positive electric charge, others a negative charge, and yet others no charge at all; some attract water molecules, others have oily affinities; some depend for their particular properties on a strategically located oxygen atom, others on a nitrogen or sulfur atom. In contrast, rarely more than four different kinds of building blocks, usually with similar properties, enter in the formation of other macromolecules.

A given protein molecule owes its particular properties to the order, or *sequence*, in which amino acids follow each other along the string. Because of the many attractions and repulsions between the characteristic chemical groups carried by the amino acids, the string most often folds into a complex ball, in which certain groups distant from each other in the string join on the surface of the ball into highly specific three-dimensional configurations. This is how the binding sites and catalytic centers of enzymes are formed. Some protein molecules retain their linear conformation and assemble into fibers, trellises, plates, and other structures. These proteins, many of which have no catalytic activity, play a structural role.

An important aspect of proteins is that the existing molecules represent a vanishingly small fraction of those that are possible. In technical terms, they occupy a vanishingly small part of the *sequence space*, which, in fact, exceeds by far anything that can materially exist, or even be imagined.[11] This fact is often brandished by creationists and other adversaries of a naturalistic explanation of the origin of life, as proof that some intelligent choice presided over the selection of the proteins present in living organisms. I shall come back to this point in a subsequent chapter.

Many enzymes act with the collaboration of specific small organic molecules, called *coenzymes*, which often contain a vitamin as their active component. In addition, enzymes frequently bear one or more metallic elements, such as iron, copper, calcium, magnesium, manganese, molybdenum, or zinc, which play an essential role in the catalytic mechanism. These facts explain the nutritional importance of vitamins and trace elements. Note that we require vitamins because, contrary to the organisms from which we obtain these substances, we lack one or more enzymes needed for their synthesis. These are special cases of metabolic errors, which we correct by an appropriate diet and partly, also, with the help of bacteria present in our digestive tract, where these microorganisms manufacture some of the vitamins we need.

A few biological catalysts do not belong to the group of proteins, but to that of ribonucleic acids (RNA). Although few, these catalytic RNAs, which are called *ribozymes*, carry out important functions, some of which probably played a crucial role in the development of life. We shall come back to them.

SELF-ASSEMBLY

Until now, we have only examined the basic conditions that allow living beings to function as chemical factories. But a collection of molecules is a far cry from a living cell, just as a heap of bricks, boards, and tiles hardly makes a house. It remains for these pieces to be combined into walls, doors, windows, a roof, and other parts, according to a definite plan. Similarly, in biological constructions, the products of syntheses have to be assembled into structural elements, such as membranes, fibers, or granules, which must themselves be combined into more elaborate structures, up to the formation of that highly complex organism, a living cell (not counting the association of the cells themselves into pluricellular organisms). In the building of a house, the construction is done by workers, following a blueprint drawn by an architect. In the building of a cell, where are the workers, where is the architect?

There are none. It all happens automatically, according to instructions written into the structures of the molecules involved. At a first level, the information is provided by the enzymes. These define, by the configurations of their binding sites and catalytic centers, what may be termed the manufacturing program of the living chemical factories, their catalogue of products, so to speak.

At the next level, assembly is guided by the structures of the molecules thus made. Enzymatic proteins often participate in this combinatorial game, by means of sites that are different from those involved in their catalytic properties. They thus form complex multi-enzyme systems, organized so as to carry out reaction sequences or cycles in a coordinated fashion. Many structural proteins devoid of enzymatic activity and other macromolecules also take part in the self-assembly of biological structures.

What is remarkable about these phenomena is their spontaneity. Even though several hundred parts may be involved in the assembly of a structure, it all happens without outside instruction. The location of each piece is inscribed in its shape, as with a piece of a puzzle, with, in addition, sufficient attractive forces to stabilize the combinations created by chance encounters. Just mix the pieces and allow them enough time to get together—which, at the rate of molecular collisions, rarely demands more than a few hours—add a pinch of ATP as a source of energy and, possibly, a catalyst or two, and an object as complex as a chromosome, for example, will form spontaneously, even in a test tube. It is as though a puzzle could be put together simply by shaking the pieces.

The key notion here is *complementarity* between molecular configurations that fit and interlock with each other. This phenomenon, whose importance could hardly be overestimated, governs the combination of pieces in self-assemblies. It also explains the selection of substrates by the binding sites of enzymes. It is likewise involved in the interactions between active agents, such as hormones or drugs, with their cellular receptors, and in the recognition of antigens by antibodies in immune defense phenomena. We shall meet it as forming the basis of all genetic information transfers. Complementarity is often illustrated by the relationship between lock and key, or between mortise and tenon. The image is suggestive but only partly appropriate. Biological mortises and tenons have over those of cabinet makers the advantage of being more flexible and adaptable, so that they can to some extent mold themselves on each other. In addition, they bear with them, in the form of mutual affinities, the "glue" that helps them stick together.

The chemistry involved in these phenomena is different from that catalyzed by enzymes. Instead of true molecules, consolidated by strong linkages between atoms, as arise by the action of enzymes, the products of assembly are looser associations between molecules that remain distinct and are kept together by relatively weak forces. These are often electrostatic attractions, such as exist between entities bearing opposite electric charges. Repulsions between charges of the same sign may also be involved, keeping certain molecules or molecular parts at a distance from each other. Often also, the kind of physical phenomenon responsible for the fact that water and oil don't mix plays a role in biological assemblies. The same kinds of forces serve to stabilize the three-dimensional conformations adopted by proteins and other large, complex molecules.

THE CENTRAL ROLE OF PROTEINS

A conclusion emerges clearly from all that we have considered thus far: proteins occupy a truly central position in the organization of life. As enzymes, proteins are responsible for the vast majority of chemical reactions that take place in living cells, including such vital processes as the construction of biological constituents, the interconversion of materials by metabolism, and the production and utilization of biological energy. In addition, proteins play a leading role in the self-assembly of biological structures of a higher order. We shall see later that most of the substances involved in regulation and in signalling are also of protein nature.

Protein molecules owe their properties to their three-dimensional shapes, which are themselves determined by the amino acid sequences of their constituent chains. This leaves a key question: how are amino acids linked to each other into protein chains according to specific sequences?

Chemically, the assembly of amino acids into proteins is carried out, like other biosynthetic mechanisms, by specific catalysts acting with the

2. What Is Life?

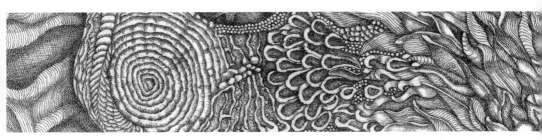

S O FAR, WE HAVE SEEN HOW LIFE PRODUCES LIFE. IT REMAINS FOR us to see how life *reproduces* life, that is, produces life similar to itself. The answer to this question is already contained in the preceding chapter. Inasmuch as the information needed to make a cell is largely written into the amino acid sequences of its proteins, all, or nearly all, that is required to reproduce the cell is to reproduce its proteins.

THE LANGUAGE OF LIFE

In principle, the simplest way to reproduce proteins would have been to use them as models for their own synthesis. This is not what happens in reality. For reasons that, as we shall see in a later chapter, tell us a great deal about the manner in which life originated, the information that guides the assembly of proteins is not provided by proteins but by nucleic acids. And these are the molecules that are actually copied. Those functions are carried out by the genetic apparatus, which, therefore, stands at the top of the hierarchy in the organization of cells.

NUCLEIC ACIDS

Thus named because they were first discovered in the cell nucleus, nucleic acids consist, like proteins, of very long chains of interconnected units. Known as *nucleotides*, these units share a common "handle" made of a five-carbon sugar, or *pentose*, and *phosphate*. To this handle is attached a nitrogenous substance, called a *base*, of which four different kinds are used. When nucleotides join to form a nucleic acid, they do so by their handles, thus creating a backbone of pentose-phosphate repeats that is the same for all the different assemblages. The bases attached to each pentose unit hang from this backbone, somewhat like those flaglets whose garlands circle used car lots or decorate the masts of ships on festive occasions.

The sequence of bases along the backbone determines the specificity of the molecules, their information content. A convenient image is to compare the chains to words and the bases to letters. Just as our words are written with an alphabet of 26 letters, we say that nucleic words are written with an alphabet of four letters, the four canonical bases. In the same imagery, protein molecules are depicted as words written with an alphabet of 20 letters, the 20 amino acids that are used throughout nature to make protein molecules. The paucity of the nucleic alphabet is compensated by the relatively greater length of the nucleic words, which generally contain many hundreds, if not thousands, of letters. For this reason, the number of distinct nucleic acid molecules that can theoretically exist (nucleic acid sequence space) is just as unimaginably immense as the number of protein molecules in the protein sequence space.[1]

There are two major kinds of nucleic acids, depending on whether the pentose is *ribose* or *deoxyribose*. They are accordingly called *ribonucleic acids* (*RNA*) or *deoxyribonucleic acids* (*DNA*). The four bases present in RNAs are adenine, uracil, guanine, and cytosine. Adenine, guanine, and cytosine are found also in DNAs, where, however, uracil is replaced by its close chemical relative, thymine. These bases are often designated by their initials.[2]

THE CIRCULATION OF GENETIC INFORMATION

In all living beings, DNA is the ultimate repository of genetic information. DNA molecules are the ones that are copied whenever the division of a cell requires doubling of its genetic information. And it is in the base sequences of DNA molecules that are written the amino acid sequences of proteins. Thus, the copying of proteins, which we have seen is the key prerequisite of reproduction, takes place by way of the copying of their DNA blueprints.

The relationship between the two sequences is simple. The amino acid sequences of proteins are dictated colinearly (in the same order) by the base sequences of the DNA molecules, each successive triplet of bases coding for a given amino acid. Such coding base triplets are termed *codons*, and the table of correspondences between codons and amino acids is called the *genetic code*.[3] The need for codons of at least three bases is evident, since there are 20 different amino acids in proteins, and only four distinct bases in DNA. Were codons made of two bases, only 16 distinct combinations would be possible, which is still insufficient. With codons of three bases, the number of combinations is 64, which is excessive. In practice, 61 of the 64 triplets are used as codons, which means that several different codons (up to six) may code for the same amino acid. The three other triplets serve as chain termination signals.

DNA does not itself direct the assembly of proteins. It does so by way of RNA molecules, appropriately termed *messenger* RNAs (mRNAs), whose base sequences are dictated by those of the corresponding DNAs. The DNA-encoded synthesis of RNA is named *transcription*, as the two alphabets are very similar (A, T, G, and C for DNA; as opposed to A, U, G, and C for RNA). The step from RNA to protein, which involves two entirely different alphabets—20 amino acids for only four bases—is understandably termed *translation*. The copying of DNA is called *replication*.

Three processes, therefore, participate in the circulation of genetic information: replication, in which the information is transferred from DNA to DNA; transcription, in which it goes from DNA to RNA; and, finally, translation, in which it moves from RNA to proteins. Note that *only information* is transferred in this way. The actual *chemical* processes involved in the making of the information-bearing DNA, RNA, and protein molecules are, like all biosynthetic mechanisms, catalyzed by specific enzymes and energetically supported by ATP. The function of the nucleic acids is to tell the synthetic machineries which of the four nucleotides or which of the 20 amino acids is to be inserted at each step in the assembly process.

The relationships just outlined are of universal significance. In all known living beings, the genetic information is stored in the base sequences of DNA molecules, reproduced by replication of this DNA, and expressed by way of the RNA and protein molecules synthesized according to the information held by the DNA. The sum total of the DNA of an organism is called its *genome*; it is subdivided into units called *genes*, each of which may be said, in rough approximation, to code for a distinct protein chain (except the few genes coding for functional RNAs, see below). The colibacillus genome contains about three million bases, that is, roughly the equivalent of ten times the number of letters in Schrödinger's "*What is Life?*" The human genome, which made headlines when its complete sequence was announced in February 2001, comprises some two thousand times this amount. For its recording—the task has, of course, devolved to a computer—about three hundred volumes of a good-size dictionary would be needed, enough to occupy some 20 meters of shelf space in a library. It is enormous; and, at the same time, it is also very little, considering that all the instructions that specify a given human being, from conception to death, are condensed in some three hundred books. These are reduced, by the marvel of molecular miniaturization, to about two meters of DNA, coiled inside a small sphere of one-hundredth of a millimeter in diameter, the cell nucleus.

A few of the RNA molecules transcribed from DNA are not translated into proteins, that is, do not serve as messengers. They play a functional role, notably in protein synthesis and in certain forms of RNA processing. These functional RNAs are the catalytic RNAs, or *ribozymes*, mentioned in the preceding chapter. Thus, the information inscribed in DNA directs all cellular activities, mostly through the proteins synthesized by the translation of messenger RNAs, and for a small part by way of functional RNAs.

LIFE'S ROSETTA STONE

Bearing the same message carved in three different languages, the Rosetta stone helped the French Egyptologist Jean-François Champollion to decipher the Egyptian hieroglyphs in the early nineteenth century. A decoding feat of incomparably greater consequence was accomplished in 1953, when the American James D. Watson and the Englishman Francis Crick discovered the double-helical structure of DNA, possibly the greatest breakthrough ever made in the understanding of life. Adopted as a logo by numerous scientific institutes and biotechnology companies, the "double helix" has become the symbol of the biological revolution of the second half of the twentieth century. In reality, the helicoidal shape of the structure is, in itself, unimportant, a simple consequence of the fact that DNA chains are naturally twisted and thus wind spirally around each other when they join. The truly significant aspect of the discovery is the "double" part, the fact that two DNA chains are associated and, especially, the reason for this association. The two chains, as brilliantly surmised by the famous team, are mutually *complementary*.

The first hint that this may be so came from analytical measurements carried out in the late 1940s by the American chemist of Austrian origin Erwin Chargaff, who worked out techniques for determining each of the four bases present in DNA. Applying these techniques to DNA samples of diverse origins, Chargaff found that the relative content of A was always equal, within measurement errors, to that of T, and that of G to that of C. With the characteristic caution of a chemist, Chargaff contented himself with recording this fact as odd and possibly significant, leaving it to Watson and Crick to make the imaginative leap from numbers to structure. Knowing, from the work of physicists using a technique known as X-ray crystallography, that DNA fibers were probably made of two helically coiled strands, the two investigators put physics and chemistry together and proposed that the two strands of the double helix are held together by their bases, with A in one of the two strands always joined to T in the other, and with G and C always similarly joined.[4] This hypothesis thus accounted in one shot for the equalities discovered by Chargaff and for the double-stranded structure of DNA.

Watson and Crick were able to bolster their proposal by means of scale models of the molecules involved. They found that A and T, on one hand, and G and C, on the other, indeed have complementary shapes, fitting each other like two pieces of a puzzle. This image is particularly appropriate, as the bases really are shaped like flat pieces with interlocking edges, with the additional participation of weak electrostatic attraction forces that stabilize their association. Called *base pairing*, this association is the most far-reaching example of the phenomenon of structural complementarity mentioned in the preceding chapter and illustrated by the lock-and-key or mortise-and-tenon relationship.

Base pairing is also involved in the structure of RNA, except that the base complementary to A is U, not T. Physically, this amounts to the same thing, since T and U have exactly the same profile in the part of their edge where they join with A. Contrary to DNA, however, RNA exists only exceptionally in double-stranded form (in some rare viruses), because RNAs most often arise as single strands, transcribed from only one of the two DNA strands. Nevertheless, because of the frequent presence of short sequences complementary to sequences located further along the same strand, RNA chains fold into numerous loops cemented by short, helicoidal, double-stranded segments. This creates bizarre, tormented shapes, which allow RNA molecules to accomplish a much wider gamut of functions than the uniformly shaped fibers of double-stranded DNA. In particular, a number of RNA molecules display catalytic properties of major importance, as we shall see. These ribozymes, already mentioned in the preceding chapter, are the only biological catalysts that are not of protein nature.

An important feature of the Watson-Crick double helix is that its two strands contain the *same information* in what may be termed, borrowing the expression from photography, positive and negative forms. Knowing the sequence of one strand, one can readily derive the sequence of the other, simply by applying the magic base-pairing formula: A = T ; G = C. For example, if one strand should contain the sequence A-G-T-G-C-A-G, one can deduce that the other strand must have (in antiparallel fashion, see note four, the complementary sequence T-C-A-C-G-T-C. This fact does not just explain the regular double-stranded structure of DNA; it has a much more profound significance.

As Watson and Crick already suggested in their original paper, in a sentence coyly introduced by the expression "it did not escape our notice," base pairing accounts for DNA replication; it is the mechanism whereby an existing DNA strand, serving as template, commands the insertion of building blocks (nucleotides) in a newly forming strand. At each step, the base presented by the template to the DNA-synthesizing machinery imposes, by pairing, the choice of the nucleotide containing the complementary base among the four that are available. If the base presented by the template is A, T is added; if it is T, A is added. Similarly G commands the selection of C, and reciprocally. In other words, instead of copying DNA as would a Xerox machine, cells construct positives on negatives, and vice versa, as in photography. With each of the two strands of a DNA double helix serving as template, the end result of the process is two identical double helices. Thanks to highly elaborate "proofreading" mechanisms, this process is of incredible fidelity: one wrongly inserted base in about one billion. It is as though a typist copied the whole *Concise Oxford English Dictionary* more than one hundred times, making only one mistake! Note that this still amounts to about a half dozen mistakes every time a human cell replicates its genome before dividing. We shall see the significance of these mistakes for evolution later.

The transcription of DNA into RNA takes place in the same way, except that the newly formed chain contains ribose (the R of RNA) instead of deoxyribose, and that the base inserted opposite A is U, not T. If, for example, the DNA sequence considered above, A-G-T-G-C-A-G, serves as template in the assembly of RNA (transcription), the resulting RNA sequence will be U-C-A-C-G-U-C, complementary, in RNA language, to that of the template.

Similar reactions can occur also with RNA as template. In RNA replication, A in the template commands the insertion of U, and U that of A, whereas in the reaction known as *reverse transcription* (synthesis of DNA on an RNA template), A commands the insertion of T, and U that of A. In all cases, G and C code mutually for each other. For example, with the sequence U-C-A-C-G-U-C as template, the product of replication will be A-G-U-G-C-A-G, and that of reverse transcription A-G-T-G-C-A-G. These reactions do not take place in normal cells; but they occur in cells infected by certain *viruses*.

Viruses are infectious, submicroscopic entities that are incapable of independent life but can enter certain cells, where the virus particles are reproduced with the help of local machineries. The simplest kind of virus consists of a small genome encased within a protein coat. The function of the coat is to bind to the target cell surface in such a way that the genome slips into the cell. This genome codes for the coat proteins and, sometimes, for an enzyme. Once inside a cell, the viral genome becomes introduced into the cell's genetic circuits, with the consequence that the viral proteins are synthesized and the viral genome is replicated. Together, these two components self-assemble into new viral particles, which leave the attacked cell, often after the latter's destruction, to invade new cells and continue multiplying and perpetrating their ravages. Viruses may be viewed as cell parasites reduced to their simplest expression; they borrow everything from the cell they parasitize, adding only what is needed to ensure their own faithful reproduction.[5]

Many viruses have a genome made of DNA like that of the cells they invade. But some have an RNA genome. Replication of the latter is accomplished by enzymes that do not exist in the host cell but are synthesized by it following instructions provided by the viral genome. There are two kinds of such enzymes. Some catalyze the direct replication of the viral RNA. As an example, the virus that causes poliomyelitis relies on such an enzyme for its multiplication. In other instances, the RNA is replicated indirectly, by way of DNA. The RNA is first reverse-transcribed into DNA, which is then transcribed back into RNA. Among those retroviruses, as they are called, are several carcinogenic agents and the infamous HIV virus, which causes the deadly acquired immunodeficiency syndrome (AIDS).

In summary, the base-pairing rules derived from the Chargaff equalities—A = T or U; and G = C—account for all biological transfers of information between nucleic acids. Whether DNA or RNA is made, whether DNA or RNA serves as the template, positives are, by base pairing, always assembled on negatives, or vice versa. The whole of genetic continuity in the entire living world is governed by two almost absurdly simple rules. Nothing could illustrate better the awesome power of structural complementarity. But this is not all. Base pairing also plays a determinant role in the mechanisms whereby, in protein synthesis, the base sequences of messenger RNAs specify the amino acid sequences of the proteins being assembled.

THE ASSEMBLY OF PROTEINS

Amino acids have in common a basic molecular skeleton characterized by the two groups, amine and acid, to which they owe their name. To this common skeleton is attached a specific group, which is different for each of the 20 amino acids that are used universally for the synthesis of proteins. In the long chains that make up proteins, the amino acids are linked by bonds, known as *peptide* bonds, in which the acid group of one molecule joins with the amine group of its neighbor, the union being stabilized by the loss of a water molecule. In the synthesis of such assemblages, the amino acids are, one by one, hooked on to the growing chain.

This process takes place in all living cells on small particles called *ribosomes*. These are tiny egg-shaped entities, about twenty-millionths of a millimeter in mean diameter, consisting of roughly equal amounts of protein and RNA. All ribosomes in a cell are the same. Their main function is to seal peptide bonds between amino acids. In this function, they have no preference. Presented with any kind of amino acid, they will hook it on, irrespective of its nature, if it is provided with the required energy (activated), and offered in the right orientation. It is interesting and, as will be seen, highly significant that this activity is catalyzed by a *ribosomal RNA* (rRNA), not by a ribosomal protein. This is one of the rare known cases of ribozyme involvement.

The nature of the amino acid to be attached to the growing protein chain at each step of the assembly process is dictated by *messenger RNAs*, which are themselves transcribed from DNA genes. The messenger RNA runs through the ribosome, much like a tape through a cassette player. The message is read, as already mentioned, by triplets of bases, or *codons*, each codon corresponding to a given amino acid according to the genetic code. As is done with a tape by a cassette player, but more jerkily, the ribosome moves the messenger RNA codon by codon, with a shift such that each codon arrives in turn at the catalytic site of the ribosome, where a peptide bond is formed between an appropriately presented amino acid and the growing chain. There, the codon dictates which of the 20 available amino acids is to be attached to the growing chain by the ribosomal machinery.

The choice of amino acids by the codons is not made by a direct interaction between the two entities. It occurs *indirectly* by way of special RNA molecules called *transfer RNAs* (tRNAs). The function of these molecules is to carry the amino acids and to bring them to the catalytic site on the ribosome, ready to be attached to the growing protein chain. Some 70 to 80 nucleotides long, tRNAs have a typical cloverleaf structure. The carried amino acid is attached to the end of the molecule corresponding to the leaf's stem. At the other end of the structure, forming what appears like the curved edge of the middle leaflet, there is a specific triplet of bases complementary to a codon characteristic of the carried amino acid and called *anticodon* for this reason. It is this anticodon, not the carried amino acid, that is recognized by the codon of the messenger RNA. A transfer RNA loaded with its amino acid can be positioned on the ribosome in a configuration that allows the amino acid to be attached to the growing protein chain only if its anticodon fits, by base pairing, with the codon presented by the messenger RNA on the ribosome. This is how amino acids are chosen, by the mediation of the anticodons of the transfer RNAs on which they are carried.[6]

A key feature of this mechanism is that the recognition step whereby the amino acid is selected occurs entirely in RNA language, following base-pairing rules. The messenger RNA codon does not "see" the amino acid itself, but only the anticodon of that amino acid's transfer RNA.

This is so true that if an amino acid is chemically converted to another after it has been attached to its specific transfer RNA, the synthesizing machinery will be fooled; it will wrongly add the modified amino acid, obeying only the information provided by the transfer RNA anticodon.

For the reason just stated, the correctness of protein assembly depends critically on the fidelity of the mechanism whereby amino acids are attached to the appropriate transfer RNAs. The selection of these two components is carried out by the enzymes that catalyze their bonding. These enzymes fish out from the metabolic pool, by means of specific binding sites, "their" amino acid and "their" transfer RNA and join them together with the help of ATP, thereby at the same time ensuring the energetic activation that was mentioned above. These enzymes, of which there are exactly 20 different kinds, one per amino acid, are the only "bilingual" entities in the machineries of life. Each of them contains, written into the binding sites that specifically select the amino acid and the transfer RNA, one line of the genetic dictionary.[7] They are collectively responsible for the actual translation from RNA language (anticodon) to protein language (amino acid), all other operations being performed exclusively in RNA (or DNA) language. The fidelity of translation depends crucially on this specificity. If one of the enzymes involved makes a "mistake," picks the wrong amino acid or the wrong transfer RNA, there will be an error in the sequence of the protein chain. It must be added that the fidelity of translation is further contingent on the exactitude of anticodon-codon pairing. Mistakes of this kind, although relatively frequent, are not as harmful as one might suspect. They do not prevent a large majority of correct molecules from being made. Furthermore, cells have developed remarkable mechanisms for recognizing and eliminating faultily assembled proteins.

THE ENIGMA OF SPLIT GENES

In 1977, two investigators working independently in two different American laboratories, the American Phillip Sharp and the Englishman Richard Roberts, made one of the most stunningly unexpected discoveries ever made in biology. Never mind what they were investigating. What they found had nothing to do with their expectations and appeared almost unbelievable at first sight: many genes do not, contrary to a view considered so self-evident that its opposite was not even envisaged, contain uninterrupted blueprints of proteins. They consist of pieces, called *exons* because they are expressed, separated by intermediary pieces, called *introns* for this reason, which are discarded. Imagine the preceding sentence written: "They consist of pi*xgtrxjiod*eces, called exons because they are expressed, separated by inter*frqnkrwkyaeix*mediary pieces, cal*utyoavbm*led introns for this reason, which are dis*gtyhncsq*carded." Such split genes are integrally transcribed into premessenger RNAs, which are then processed in such a way that the introns (italicized) are excised and the exons spliced together into the final message; somewhat like the way films are edited in the cutting room. One remains perplexed by this complication, which obviously introduces additional sources of errors into the genetic machinery. That gene splicing exists clearly indicates that it plays, or has played, a highly useful role. Otherwise, one would expect it to be eliminated by natural selection.[8]

On the contrary, there are good indications that split genes may have something to do with vertical evolution toward increasing complexity. Almost totally absent in bacteria and protists, split genes are present in pluricellular plants and animals, being more numerous in the so-called higher organisms than in the lower ones. Scientists are, however, deeply divided on the interpretation to be given to this fact. Interestingly, several small RNA molecules participate in the cutting and splicing of RNA transcribed from split genes. This, next to protein synthesis, is another major instance of ribozyme involvement.

THE "CENTRAL DOGMA"

All the discoveries of modern biology show that proteins can be reproduced only by way of the corresponding nucleic acids. Direct copying of a protein molecule has never been observed. Neither has the reverse translation of a protein sequence into a nucleic acid sequence. The transfer of information between nucleic acids and proteins is strictly *one way*. This is what Crick has called the "central dogma." One would prefer the term "postulate" or "rule." Dogmas have nothing to do with science. In any event, Crick's central dogma does seem to have the character of a nontransgressable law.

A consequence of the central dogma is that an acquired modification of a protein cannot be transmitted to offspring, because the information cannot be transferred to the gene coding for that protein. This, as we shall see in Chapter 7, is a major reason for rejecting Lamarck's theory of the inheritance of acquired characters. Although undoubtedly correct in the vast majority of cases, this inference may not be entirely valid. It is possible that certain shapes are hereditarily transmitted by contact, rather than by genes.

CONTINUITY OF SHAPES

We have seen in the preceding chapter the role played by spontaneous self-assembly in the formation of cellular structures. Note, however, that the cellular factories do not build copies of themselves next to themselves or in a separate compartment. They grow harmoniously and then divide. So do many intracellular entities. It may happen, in such cases, that existing structures affect the manner in which certain self-assembly processes take place. Membranes, those tenuous films that envelop all cells and partition many into numerous distinct compartments, are a characteristic example.

Biological membranes are constructed from proteins and from special kinds of lipids (fatty substances). Leaving aside the nature of the molecules involved, the important point is that such structures never arise *de novo;* their assembly always takes place by *accretion*, that is, by the insertion of additional materials into a pre-existing membrane. In this process, some materials are taken up because they fit into the existing structure; others are left out because they don't fit. The membrane thus selects the inserted materials. Just as there is a continuity from cell to cell, going back to the first cells ancestral to all forms of life, there is a similar continuity, from membrane to membrane, to the membranes of those primitive cells. This has led the German-American biologist Günter Blobel to paraphrase Virchow's aphorism, *omnis cellula e cellula* (Chapter 1), by: *omnis membrana e membrana*, all membranes come from membranes. In view of these facts, the possibility that changes in membrane shape acquired during the life of a cell may be transmitted to the cell's descendants deserves to be contemplated.

A more subtle kind of shape transmission has been highlighted by the discovery of *prions*, which are infective agents responsible for several grave diseases, including bovine spongiform encephalitis (BSE), or mad cow disease, and its human counterpart, Creutzfeldt-Jakob disease. BSE made headlines when it broke out in Great Britain in the early 1990s, because of the risk of its transmission to human subjects. As this book is being written, the threat has, if anything, become greater and caused drastic measures to be decided in Europe. Prions were first taken for viruses, until the

American investigator Stanley Prusiner showed them to be made exclusively of proteins, without the accompaniment, invariably found in viruses, of nucleic acids providing the information for their reproduction.

Proteins that are reproduced without nucleic acids! A major heresy is suggested, a faulting of the central dogma. Things, fortunately, are not so bad. It is now known that prion proteins exist normally in the organisms they infect, where they are reproduced in a perfectly orthodox fashion, by the expression of local genes. But what is reproduced in this way is only, as for all proteins, the amino acid sequence of the molecule. What is changed in the prion, compared to the normal protein, is the three-dimensional conformation the amino acid chain adopts in folding. And what makes a prion infectious is that the wrongly folded protein is exceptionally resistant to factors such as heat, dessiccation, or enzymatic degradation, and that it can confer its defective shape to the normal protein by contact. The abnormal protein deposits responsible for the grave cerebral lesions characteristic of prion diseases are believed to arise in this way. This, at least, is the explanation defended by Prusiner, which, vigorously contested at first, is now widely accepted.

It is conceivable that the phenomenon disclosed by pathology may have a physiological counterpart and that certain normal protein conformations also may be transmitted by contact. The future will tell. At present, the possible importance of hereditary transmission by continuity of shapes cannot yet be assessed. Neither is it known to what extent accidental modifications of shapes may be similarly transmitted. But such possibilities deserve to be kept in mind.

Lessons of Life

At the end of this, perforce, highly condensed and simplified description of the main properties of life, three conclusions emerge. First, *life is one*. Already made clear in the Introduction, this affirmation is reinforced by all that has been seen. All the known living beings that subsist, grow, and reproduce on this planet—the trees and the flowers, the fungi and the mushrooms, the extraordinary richness of animal life, in the waters, in the air, and on land, including human beings, together with the immensely varied world of invisible bacteria and protists—all maintain and propagate themselves by the same mechanisms, no doubt inherited from a common ancestral form. The revelation is awe-inspiring. So is the realization that the unrelenting human urge to understand has, just in our times, disclosed life's secrets for us.

Second major conclusion: *life is chemistry*, to which must be added physics to the extent that physical chemistry is involved in such phenomena as nerve conductance or membrane potentials. This point has already been made early in the preceding chapter. All that followed has but strengthened it. Our explanations of life invariably call on molecular structures and interactions. The language of life is the language of biochemistry.

This truth tends to be overshadowed nowadays by the advances of genetics and molecular biology. The language of genetics is so appealing in its simplicity, so easily accessible to the layperson, that the realities behind it are no longer always taken into consideration. Many practitioners of molecular or evolutionary biology pursue their activities without calling on biochemical concepts, of which they are sometimes surprisingly ignorant. In their computer simulations, theoretical biologists replace molecular structures by symbols, and chemical reactions by algorithms. Such exercises can be useful and illuminating. But to call their outcome "artificial life" is misleading. If life is ever created artificially, it will be in a test tube, not in a computer.

This point will become evident when we consider the origin of life. Just as we cannot possibly understand life without chemistry, we must perforce look at its origin in terms of chemistry. In this chapter, I have done my best to avoid technicalities, so as to reach the largest possible number of readers, while trying, nevertheless, not to cross the boundary beyond which simplification becomes misrepresentation. Those readers who still found the going rough are encouraged to renew or improve their acquaintance with chemistry. Twenty-first-century culture mandates a minimum of chemical literacy. An elementary initiation to this discipline has become indispensable and should be part of the cultural assets of every individual.

The third lesson we can draw from this survey concerns the *central role of RNA* and related nucleotide derivatives, in particular ATP, in the common blueprint of life. This is particularly true of the synthesis of proteins, where all the main functions are carried out by RNA molecules: ribosomal RNAs, transfer RNAs, and messenger RNAs, not counting the small RNAs involved in gene splicing. To be sure, many proteins are implicated as well, in particular the "bilingual" enzymes that attach amino acids to transfer RNAs. Nevertheless, the central role of RNAs in protein synthesis suggests strongly that, under the primitive conditions when proteins were only starting to be made, the machinery involved may have consisted exclusively of RNA molecules. This point will be discussed in greater detail in Chapters 4 and 5. But before we get to this question, we must first take a look at the cosmic and planetary conditions that formed the cradle of life.

3. Where Does Life Come From?

WHAT WAS THE ancestral form from which all known living beings descend? When did it appear? Where did it come from? At the time of my youth, the possibility of answering those questions was so remote that very few biologists bothered to ask them. Vitalists, of one ilk or another, felt the questions to be unanswerable by science. Even those biologists, probably a majority, who believed that life must have arisen spontaneously by purely natural phenomena, theoretically accessible to research, mostly considered these phenomena unknowable in the existing state of science and not worth investigating. I remember hearing, at the Third International Congress of Biochemistry held in Brussels in 1955, a lecture by the acknowledged pioneer of origin-of-life studies, the Soviet biochemist Alexander Oparin, whose book titled *The Origin of Life on Earth* was first published in 1924. Prefaced by the ritual homage to Stalin, his exposition struck me as laughable, if not suspect of sinister, Marxist connotations. It seemed to me futile to look for the origin of something of which almost nothing was understood.

Things have greatly changed. Today, hundreds of distinguished investigators devote their efforts to the origin of life. They have their own society, congresses, and journals. Some of their books are bestsellers. The domain has ceased to be ridiculous. On the contrary, it has become, thanks to many discoveries, thanks, especially, to immense advances in our understanding of life, one of the most exciting research topics of our time.

THE LAST COMMON ANCESTOR OF ALL LIFE CAN BE RECONSTRUCTED FROM ITS DESCENDANTS

In Chapter 1, life was defined as what is common to all living beings. At first sight, one would expect this definition to apply almost unchanged to the last common ancestor of all life on Earth, from which all those shared properties presumably were inherited. Things, however, are not so simple; and the properties of the last universal common ancestor (LUCA) have become a subject of intense speculation and discussion.

The main difficulty comes from the possibility that certain genes may have arisen later in a given evolutionary line and subsequently entered the other lines by *horizontal transfer*, that is, transfer between different species (vertical transfer being that occurring within a species by the normal mechanisms of heredity). Present evidence suggests that this is a widespread phenomenon in the bacterial world. Some genes transferred in this way, although present in all living beings, could have been absent in the LUCA. This possibility is real but can only apply to the very early days of life, when few species existed and they occupied the same environment. A species isolated from the others could no longer receive genes from them. Its progeny would lack the genes in question. Indeed, to be common to all forms of life, properties acquired by horizontal gene transfer must have been gained by all the lineages that have left descendants until our time.

The same objection applies to the possibility that certain genes that were not present in the LUCA arose later, in separate lines, by *convergent evolution*. Such a phenomenon, if it occurred at all, could obviously be of significance only when very few distinct evolutionary lines still existed. The more numerous the lines, the smaller the probability of convergent evolution endowing all with the same gene.

A more likely possibility is that certain genes that were present in the LUCA were subsequently lost, during evolution, by a number of lines. Thus, the absence of a property in some present-day organisms in no way proves that this property did not exist in the LUCA. This is an important point, but it leads only to our underestimating the properties of the LUCA. Taking only those properties that are common to all known living beings, we have a pretty comprehensive picture of the LUCA. The only caveat is that the LUCA was, perhaps, not a single organism, but, as has been suggested, a collection of organisms sharing a common pool of genes that were freely exchanged by horizontal transfer.

With due regard to these uncertainties, we have enough information to sketch a portrait of the ancestor likely to resemble the actual ancestor fairly faithfully. First, it was manifestly a cellular organism, almost certainly unicellular and, for evident reasons of simplicity, prokaryotic rather than eukaryotic.[1] It possessed the minimum characteristic attributes of all cells, to wit a peripheral membrane, perhaps supported by an external wall; a cytoplasm, site of metabolism; and a chromosome, vehicle of heredity.

The metabolism of the common ancestor must have involved at least several hundred distinct chemical reactions, catalyzed by protein enzymes already assisted by the main coenzymes known today. These reactions must have included some of the metabolic pathways present in the great majority of extant organisms, notably certain key anaerobic fermentation pathways, as will be mentioned later. ATP no doubt was the ancestor's main energy vector. Its genes were made of DNA, which was replicated, transcribed, and translated according to the same complementarity rules and genetic codes as prevail today. Its proteins were synthesized on typical ribosomes, with the help of the three kinds of RNA now involved in this process. In short, the last common ancestor of all life on Earth may not have been very different from some present-day bacterium. Some important gaps, however, remain in this hypothetical picture.

First, did the primitive ancestor manufacture its own foodstuffs or did it derive them from outside? In technical terms, was it autotrophic or heterotrophic? At first sight, one would expect it to be self-sufficient. But this is not necessarily so. We shall see that the primitive Earth may have been abundantly supplied with organic substances of nonliving origin, from which the first forms of life are suspected to have arisen. Therefore, these forms may well have been heterotrophic, feeding on those substances or, alternatively, on coexisting autotrophic forms that subsequently disappeared without leaving descendants. However, such situations, if they ever existed, can only have been temporary. Autotrophy must necessarily have developed in a stable form before available food supplies were exhausted. Otherwise, life would have become extinct. The question is whether the primitive ancestor already was self-sufficient, or whether autotrophy arose later in some of its descendants. Scientists remain divided on this issue.

Another question concerns the energy source exploited by the primitive ancestor. One can rule out oxidation, whether of organic or mineral substances, as all the available evidence points to the atmosphere of the primitive Earth as containing little or no oxygen, which is a by-product of biological photosynthesis. The primitive ancestor, therefore, was adapted to life without oxygen—it was anaerobic—and it most likely depended on the kind of fermentations, such as the conversion of sugar to alcohol or lactic acid, that sustain anaerobic life today. This hypothesis is all the more plausible because fermentation systems exist in the great majority of living beings and involve energy-retrieval mechanisms that are simpler than the oxidative processes.

A heterotrophic organism could have subsisted on such reactions. But we have seen that the primitive ancestor may well have been autotrophic. Here we have a choice between two known forms of autotrophy: photosynthesis, which derives its energy from light, and chemotrophy, which depends on mineral chemical reactions. The latter are mainly oxidations, but some exist that do not require oxygen, for example, the formation of methane from carbon dioxide and hydrogen. Several hydrogen-generating processes are known that could have taken place in the oceans or in the soil of our young planet and could have supported such a metabolism.

This leads to a last question, closely related to the preceding one: what type of environment was occupied by the last common ancestor? The organism most likely lived in water. But at what temperature? At the surface or deep down? If it was photosynthetic, its habitat must perforce have been on the surface and, hence, temperate. However, there has been considerable interest lately in the possibility that life may have originated in deep, very hot waters, such as are found in volcanic geysers and, especially, in those deep-sea hydrothermal vents (black smokers) that spew high-pressure jets of overheated water, laden with mineral elements, through cracks opening at the bottom of oceans. In the last few years, these sites have been found to harbor a number of strange bacteria adapted to very high temperatures, sometimes exceeding 100° C. According to molecular sequencing studies, these organisms are among the most ancient known. We shall see later the possible significance of these findings (see Chapter 8).

A striking feature of our reconstructed portrait of the primitive ancestor is its modern character. Should this organism be encountered today, it might well not betray its immense antiquity, except by its DNA sequences. It must necessarily have been preceded by more rudimentary forms, intermediate stages in the genesis of the elaborate structural, metabolic, energetic, and genetic systems shared by all present-day living beings. Unfortunately, these forms have left no similarly primitive descendants that would allow their characterization. This lack greatly complicates the problem of the origin of life.[2]

Life Appeared on Earth Nearly Four Billion Years Ago

The Earth was born about 4.55 billion years ago. It condensed, together with the other planets of the solar system, within a disk of gas and dust whirling around a young star that was to become our Sun. Phenomena of extreme violence, incompatible with the maintenance of any sort of life, surrounded this birth. For at least a half billion years, comets and asteroids battered the forming Earth, rendering it incapable of harboring life during all that time. Some impacts may even have been sufficiently violent to cause the loss of all terrestrial water by vaporization, following which the oceans would have been replenished with water brought down by comets. According to this version of events, present oceans would date back to the last wave of intense cometary bombardment, which experts believe took place some four billion years ago. There are signs that life was present on Earth soon after these cataclysms came to an end.

Fossilized remnants of typical bacteria (microfossils) and, even, of complex bacterial colonies, called stromatolites, astonishingly similar to extant living formations, have been found in a number of ancient rocks, including some Australian cherts estimated to be almost 3.5 billion years old. According to their discoverer, the American microfossil expert William Schopf, the Australian traces originate from highly evolved bacteria, closely related to present-day cyanobacteria, that is, bacteria that carry out a sophisticated kind of oxygen-generating photosynthesis. This claim, which, as we shall see in Chapter 8, raises some difficulties, has recently been seriously questioned.[3] There remains, however, a distinct possibility—many would say a strong likelihood—that some forms of life were already present on Earth 3.5 billion years ago, perhaps even earlier. This is indicated by the finding, in certain ancient carbon deposits, of what is generally interpreted as an atomic signature of biological activity, that is, an excess of the light carbon isotope, ^{12}C, over its heavy isotope, ^{13}C.[4] This clue has been detected, in Greenland, in rocks that are 3.85 billion years old (and also in the Australian traces referred to previously). Life could even be more ancient. We would be unable to know, as any trace it might have left could not have been preserved until our days.

Some investigators believe that the time elapsed between the moment when Earth became livable and when life appeared was too short for something as complex as a living cell to emerge. Hence the hypothesis that life came from elsewhere. What are we to make of it?

Did Life Come from Outer Space?

The notion that life is of extraterrestrial origin has had illustrious proponents. Among them, the Swedish chemist Svante Arrhenius, winner of the 1903 Nobel prize in chemistry and remembered today for a prophetic view of the greenhouse effect,[5] coined the term "panspermia" for his theory that germs of life exist everywhere in the cosmos and continually fall on Earth. More recently, a celebrated British astronomer, Sir Fred Hoyle, who died in 2001, has claimed, together with a Sri-Lankan colleague, Chandra Wickramasinghe, to have detected spectroscopic proof of the presence of living organisms on comets. We shall see later what this evidence is. Francis Crick, codiscoverer, with James Watson, of the double-helical structure of DNA, has even proposed, with another scientist of British origin, Leslie Orgel, that the first living organisms may have reached Earth on board a spaceship sent out by some "distant civilization." He has given the name "directed panspermia" to this hypothesis.

Leaving aside the spaceship, of which no sign has been found so far, an extraterrestrial origin of life is perfectly plausible. The often-voiced objection that living organisms could not withstand the physical conditions that prevail in space, especially the intense ultraviolet radiation, does not hold, as it is readily conceived that comets or meteorites may offer protection to the organisms. Destruction by heat upon entry into the terrestrial atmosphere could similarly be prevented. Moreover, the possibility that life may be a widespread phenomenon, existing in many sites of the universe, is increasingly being entertained. I shall examine this question in Chapter 17. Thus, the eventuality of living organisms travelling through space on various "flying objects" is far from implausible. But what about the evidence?

The argument that there was not enough time for life to arise locally on Earth rests on a purely subjective and arbitrary estimate, supported by no objective element. There is no proof that the emergence of life must have required hundreds of millions of years, as has been maintained. On the contrary, as I shall point out later, the essentially chemical and deterministic vision one must have of this phenomenon rather leads to the belief that life arose relatively fast, in a time span probably to be counted in millennia rather than in millions of years. In this view, the window of some 100 million years allowed by present data leaves more than ample time for life to have been born on Earth. It is even possible that life arose and disappeared many times before establishing itself.

There remain the many observations, clearly undeniable, showing that the elementary constituents of life exist on comets and other celestial objects. But are these substances products of life, as is believed by the defenders of panspermia? Or are they, on the contrary, the fruits of spontaneous chemical reactions? We shall see that the second explanation is considered the more probable of the two.

The Cosmos Is a Vast Laboratory of Organic Chemistry

For millennia, all that humans have learned of the Universe around them has been provided by the "pale light falling from the stars,"[6] from which, since Galileo, the growing power of telescopes has been extracting increasingly detailed information. But our eyes, even helped by the best optical instruments, perceive only a minute fraction of the radiation that comes to us. They see only radiations of wavelengths comprised between 400-(violet) and 800-(red) millionths of a millimeter. This narrow band is inserted within a huge span of invisible radiation, which, on the side of shorter wavelengths (higher energies) ranges from the ultraviolet to

X rays, γ rays, and cosmic rays, with wavelengths reaching below one-billionth of a millimeter, and, on the side of longer wavelengths (lower energies), extends from the infrared to kilometric Hertz waves. Today, the new discipline of *radioastronomy* sweeps a good part of this span by means of instruments of ever increasing sensitivity. The information gathered in this way is immensely richer than that provided by visible light alone.

The most important data are chemical. This is because substances betray their nature by the radiation they emit or absorb. Sodium, we know, emits yellow light; neon, red light. If the light emitted by a sodium lamp is decomposed with a prism, only two yellow bands are seen, instead of the usual rainbow. If, on the other hand, a ray of white light that has passed through sodium vapor is likewise decomposed, the same two bands are seen, but now in the form of black bands in the yellow region of the spectrum. Emission bands have become absorption bands. It is by this kind of analysis that hydrogen has been identified as a component of the Sun. Helium, as its name recalls (*hêlios* means sun in Greek), was even discovered first in the Sun, before being found on Earth.

What is true for visible light is true also of the radiations that escape the eye. Such radiations can be similarly decomposed by appropriate "prisms" and the spectra thus produced can be recorded and analyzed for the signature of certain atoms or molecules. The wavelength region around one centimeter is particularly rich in this respect. Microwave ovens function with this type of radiation. The waves that come to us from space could never power the tiniest of ovens, but their feeble messages can nevertheless be amplified and decoded in a detailed manner with the instruments now available. A complication in this kind of analysis comes from the atmosphere, which blurs the signals and adds its own. For example, there is so much water in the atmosphere that detecting traces of this substance elsewhere is impossible. But there are ways of getting around this; the simplest is to put the instruments above the atmosphere, on satellites or spaceships, as is increasingly done.

Spectral analysis at a distance is only one means. Robot instruments carried by spaceships have performed a number of direct measurements on comets. And, especially, it has been possible to apply all the resources of modern technology to meteorites that have fallen on the Earth. These various explorations have revealed the surprising fact that organic chemistry is the most banal and abundant chemistry in the whole universe.

Two centuries ago, the founders of chemistry designated as organic the chemistry of substances made by living organisms with the help, many believed, of a special vital force. This notion was first contradicted in 1828, when the German chemist Friedrich Wöhler synthesized urea; and it was definitively disproved in 1897, when another German chemist, Eduard Büchner, discovered that yeast juice devoid of living cells could convert sugar into alcohol. In the opinion of Pasteur, who unfortunately died two years before Büchner's results became known, this fermentation required "Life."

Since then, laboratory organic chemistry has produced spectacular developments that have fertilized industry and given us the entire gamut of modern plastic materials and synthetic fibers, an abundance of drugs, and many other so-called synthetic substances. It has become evident that organic chemistry is none other than carbon chemistry and that it owes its exceptional richness to the particular associative properties of the carbon atom.

Some sort of residual vitalistic aura has, nevertheless, persisted around organic chemistry, perceived almost subliminally as a kind of chemistry practiced only by living beings, including organic chemists. Space chemistry has shattered this last refuge of vitalism by showing that organic substances are spread throughout the cosmos, where they make up an important fraction of cold matter. Small radicals and molecules, made of only a few atoms of carbon, hydrogen, oxygen, sometimes nitrogen or sulfur, are present on minute dust particles that make up extremely tenuous clouds—more rarefied than the best vacuum we are capable of producing on Earth—but immensely extended, filling vast regions of space with what is known as *interstellar dust*. When such particles get together, the small molecules they contain interact to generate larger entities, of which many have now been identified on comets and other celestial bodies, especially meteorites, which have lent themselves to detailed analyses.

The results of these analyses are nothing less than flabbergasting. Not only have they revealed the existence of numerous organic molecules of manifestly extraterrestrial origin, but these molecules turned out to comprise many characteristic constituents of life, such as, for example, amino acids. Astonishingly, these findings have made little impact in the scientific world, even less so in the world in general. Yet, the message they broadcast is supremely important. The chemical germs of life are *banal products of space chemistry*. There is "vital dust" everywhere in the universe.

Before this conclusion can be accepted, it must be ascertained that the molecules do not come from some terrestrial contamination. Especially, the possibility that they have been manufactured by extraterrestrial living organisms must be ruled out. As far as contamination is concerned, detailed examinations have allowed this explanation to be categorically excluded in a number of cases. As to a biological origin of the substances, this, obviously, is the interpretation favored by the partisans of panspermia. The majority opinion, however, is that the molecules are of nonbiological origin. A good reason for adopting this view is that the same molecules are readily obtained in the laboratory under conditions that could have prevailed on Earth four billion years ago.

THE CHEMISTRY OF LIFE IS REPRODUCED IN A TEST TUBE

The story starts in Chicago in 1953—the year of the double helix!—in the laboratory of Harold Urey, an American physicist world-renowned for the discovery of heavy hydrogen, or deuterium. Later in his career, Urey had become interested in the origin of the planets. He had put forward the hypothesis that the atmosphere of the young Earth was very different from what it is today. It was, he believed, devoid of oxygen and rich in hydrogen and hydrogen-containing substances, such as methane (CH_4), ammonia (NH_3), and water vapor (H_2O). There is agreement among experts on the absence of oxygen, almost certainly a product of life, but the abundance of hydrogen is disputed by many. Be that as it may, a young student working in Urey's laboratory was sufficiently impressed by Urey's theory to ask the question how repeated lightning might have affected the atmosphere postulated by his mentor. Against the advice of the latter, who found the project too iffy for a doctoral thesis, the student built a glass enclosure within which the gas mixture postulated by Urey was subjected to a succession of electric discharges. The results exceeded the student's wildest dreams. In a few days' time, almost 20 percent of the methane carbon had been converted into amino acids and other typical biological constituents.

This historic experiment almost instantaneously propelled the name of the student—Stanley Miller—into the firmament of celebrities. It also inaugurated a new discipline, abiotic (without life), or prebiotic (before life), chemistry, which aims at synthesizing biological compounds under conditions that might have prevailed on Earth before the appearance of life. Many elementary constituents of life have thereby been obtained under plausible prebiotic conditions. The products of this new chemistry show remarkable similarities, both qualitative and quantitative, with substances detected in meteorites. What is reproduced in the laboratory seems close to what occurs spontaneously in space.[7]

These discoveries have returned to the foreground the possibility that life arose naturally, a possibility long discredited by the celebrated experiment done by Pasteur, which crushed poor Félix-Archimède Pouchet, a defender of spontaneous generation, in front of the entire Académie des Sciences assembled in solemn gathering.[8]

Did Life Arise Naturally?

For a large part of the general public, life arose through direct action by a Creator. Not only strict creationists, who rest on a literal interpretation of the Bible, subscribe to this opinion. So do many members of more open-minded religious groups. Even outside any religious creed, the origin of life is often viewed as an insoluble mystery, within the context of some unconscious latent vitalism. Rare are those who, being cultured but devoid of scientific grounding, picture life as having spontaneously arisen through the play of the same physical and chemical laws as rule other natural phenomena, such as the formation of planets, the shifts of the Earth's crust, tidal movements, or the erosion of mountains. Pasteur's triumph, one of the rare scientific events to have earned a place in popular history books, is perhaps not foreign to this attitude.

Yet, all that we have seen so far supports a naturalistic explanation of the origin of life. There is first the fact, related in the preceding chapters, that life has proved entirely explainable in physical-chemical terms. What is true of life now is very likely to be true also of its origin. If life functions without the help of a vital principle, as we know it does, we are entitled to assume that its birth likewise took place without the intervention of such an entity. Another encouraging fact is the discovery, just recalled, of the vast cosmic chemistry that abundantly produces amino acids and other organic substances entering into the composition of living beings. If, as seems reasonable to suppose, those substances represent the chemical seeds from which life developed, it may be said that at least the first step in the birth of life was the outcome of natural processes.

But this is only a first step in what must have been a very long succession of steps. As will be seen in the next chapters, we are mostly left with speculative hypotheses to explain the manner in which the basic building blocks provided by cosmic chemistry might have combined into larger molecules, such as proteins and, especially, nucleic acids, not counting the more complex assemblages from which the first biological structures arose. One may well wonder, therefore, whether we will ever succeed in explaining the origin of life naturally or, even, whether this phenomenon is naturally explainable.

In the view of most scientists interested in the problem, one can but answer the last question affirmatively, at least as a working hypothesis. No scientist could think otherwise, as this hypothesis represents the fundamental postulate of any scientific investigation. To assume the opposite amounts to denying the possibility of finding an explanation for the phenomenon one studies and thus declaring one's research futile. Independently of any preconceived idea, science must proceed on the assumption that the problems it approaches are soluble. There will always be time to call on "something else" after all attempts at finding a natural explanation have failed. In the case of the origin of life, this is still far from being the case.

The fact remains that, as long as the problem is not solved, the tendency to invoke "something else" will subsist. It is the attitude even of a small minority of scientists, very few in number but much publicized. According to these dissenters, there are intrinsic reasons for believing that life, as we know it, cannot possibly be the fruit of natural phenomena. Worded in apparently irreproachable scientific terms, such affirmations are enthusiastically greeted and fervently propagated, not only by traditional creationist circles, but also by diverse groups who, while claiming to accept the findings of modern biology, emphasize that "science does not explain everything" and defend the thesis, of so-called intelligent design, that detects in the properties, origin, and evolution of life the intervention of an influence other than the simple play of natural laws.

An argument brought forward in favor of this thesis by the American biochemist Michael Behe represents what he calls "irreducible complexity,"[9] a notion he defines as the state of "a single system composed of several well-matched interacting parts that contribute to the basic function, wherein the removal of any one of the parts causes the system to effectively cease functioning." This definition, which he illustrates with the "humble mousetrap," applies according to Behe to numerous biochemical systems, for example, the flagellum that propels bacteria or the enzymatic cascade that governs blood clotting.

No one will deny that these systems and many others conform to the proposed definition. One cannot remove one of their parts without impairing their functioning. But this in no way proves that, as is claimed by Behe, these systems can have arisen only with the help of an outside intelligence that adjusted the various parts according to a pre-established plan in which their role in the whole was foreseen. Such an affirmation ignores the possibility of an evolutionary process that might, with the help of natural selection, have led to increasing complexity by way of intermediary stages each of which fulfilled a useful function. Many examples of such processes are known. Thus, it is known that the principal proteins of the transparent eye lens were recruited in the course of evolution from enzymatic proteins that played an entirely different role.

Another objection frequently addressed to the theory of a natural origin of life calls on the fact, already mentioned in the preceding chapters, that life uses only an infinitesimal fraction of the possible protein or nucleic acid sequences, or, in more technical terms, occupies only an infinitesimal part of the sequence space. Remember, the number of different protein chains of 100 amino acids that can exist is 10^{130} (one followed by 130 zeros), that of possible nucleic acid chains of 300 bases 10^{180} (one followed by 180 zeros).[10] Numbers of this size exceed by far anything that can exist in reality, or even be conjured up by our imagination. What, then, of the number of sequences nascent life was able to test, in about 100 million years, with only the materials available on the surface of our planet (or, for that matter, on whatever celestial object provided the cradle of life)? Yet, the bacterial ancestor of all life must have contained hundreds of distinct such molecules, many of them longer than those being considered here. The problem thus arises as to how emerging life could, without guidance, conceivably have selected its constituents from such an immeasurably huge number of possibilities.

To explain the generation of the ancestral proteins—the fact that this process took place by way of nucleic acids makes no difference to the argument—by the natural unfolding of chemical processes, one would have to assume either that almost any random combination of amino acids will produce a collection of proteins adequate to make a viable cell, or that the molecular specificity of the processes involved was such as to almost obligatorily produce the right mixture. The first explanation is ruled out by what we know of biology, which tells us that the functions of proteins often are exquisitely dependent on specific sequences, to the point of being frequently impaired by the replacement of a single amino acid by another. The second explanation is ruled out by what we know of chemistry. Processes of the required precision simply do not take place. Hence, it is claimed, there must have been "something else." Such is the conclusion arrived at in a solidly argued book by the American mathematician William Dembski significantly titled *The Design Inference.*[11]

Here, again, the argument neglects the historical dimension of these phenomena. As will be mentioned later (Chapter 5), there are good reasons for believing that the first sequences were much shorter than today's and that nascent life has reached its present position in the sequence space by a gradual pathway, each stage of which, honed by natural selection, allowed extensive exploration of the available sequence space. Intervention by a directing intelligence is not mandatory.

Contrary to what is sometimes claimed, a naturalistic view of the origin of life does not necessarily exclude belief in a Creator. The notion, propagated at the same time, though for opposite reasons, by militant atheistic scientists and by many antiscientific circles, that the findings of science are incompatible with the existence of a Creator is false. But these findings at least call for a revision of the image one makes of this Creator. It cannot be a God who, according to the familiar animist saying, "blew life into matter." This notion is no longer valid now that we know that there is no such thing as a vital principle. Likewise, there is every reason to believe that the elementary constituents of life form spontaneously in many parts of the universe, by the sole operation of physical-chemical phenomena. Thus, if we wish to call on some creative act to explain the origin of life, we are led to imagine a God who got into the act at some precise moment, forcing the molecules of basic constituents to interact against their natural tendency until a machinery capable of functioning under its own steam had been built, following which He would withdraw from the game and allow the sole physical and chemical forces to play freely. This naïve picture of a divine engineer interfering just enough with the laws of his creation to achieve an objective looks too much like a contrived, *ad hoc* hypothesis to be intellectually acceptable. Why not imagine a God who, from the start, created a world capable of giving rise to life by the sole unfolding of natural laws of His own devising?

This view, as we shall see in the last chapter, is now defended by many deists, including a number of scientists.

Is Life the Product of Chance?

While scientists generally agree to attribute the origin of life to natural phenomena, the degree of likelihood of these phenomena is very diversely appreciated. According to many scientists, among them some of the most illustrious, life is the product of highly improbable events that are very unlikely ever to occur anywhere else and that could very well not have happened on Earth were it not for an extraordinary combination of circumstances. Any failure to reproduce the phenomenon in the laboratory is thus explained beforehand. It is pointed out that highly improbable events take place all the time without our according them any attention unless there is something special about them. Thus, in the game of bridge, each distribution of the 52 cards among the four players has one chance in 5×10^{28}, that is, in 50 billion billion billion, of being dealt. This guarantees with near certainty that each distribution is a unique event that never occurred previously and will never occur again in a foreseeable future. Nevertheless, bridge players do not spend their time marvelling at their cards with the feeling, at each deal, of being witness to an extraordinary event. They would do so only if there should be something uncommon about the distribution. If, for example, the 13 spades, hearts, diamonds, and clubs should each be gathered in a single hand, the event would cause a sensation, and the whole world would be apprised of it by bridge columnists. And yet, this distribution is no more improbable than any other.

Such, it is claimed, could be the case also with life. As with a bridge deal in which each hand contains a complete suit, the first living system could have been, among innumerable other arrangements of matter of equally low probability but of no particular interest, the outcome of an extremely improbable combination of circumstances, so improbable that it is virtually certain to be unique. In this view, life appears as a cosmic accident devoid of significance. In the words of the late French biologist Jacques Monod, "the Universe was not pregnant with life."[12]

Such a conception is acceptable provided the stroke of luck concerns a single event. It could be an extremely improbable event, but there can be only one. Indeed, from the moment several highly improbable events are required to reach a certain goal, the probability of ever getting there soon approaches zero, since the probability of a complex series of events is the product of the probabilities of its individual steps. Thus, the probability of the same bridge distribution being dealt were it only twice in succession is $(5 \times 10^{28})^2$, that is one chance in 25 followed by 56 zeros; that and zero, practically speaking, amount to the same thing.

It is obvious that life cannot possibly have arisen in one shot. For this to have happened, nothing short of a miracle would have been needed. The process, if it took place naturally, must by necessity have been composed of many steps, most of which, as we have just seen, must have had a high probability of taking place. Thus, the "lucky chance" hypothesis implies that a singular event of extremely low probability occurred in a series in which the great majority of the steps that came before and after followed a highly deterministic course, imposed by the prevailing conditions. Once again, we are faced with a possibility that cannot be ruled out but is hardly conceivable in realistic terms. From what we know of life, it is difficult to see how it could have developed by the succession of a very large number of spontaneous events, broken by a single barrier that could have been surmounted only with an extraordinary assistance of chance. Starting from the basic constituents provided by space chemistry, life must have arisen through a complex fabric of interconnected reactions involving a large number of different substances. This development no doubt relied on numerous discrete events, but not on a single event of extremely low probability.

Another reason for ruling out a critical intervention by chance in the development of life is that *chemical* processes were involved. Chemistry deals with strictly *deterministic*, reproducible phenomena that depend on the statistical behavior of trillions of molecules of various kinds. Were it not so, there would be no chemical laboratories, no chemical industries. When substances A and B are mixed under specified conditions, the outcome is always C. If a student fails to get C in the laboratory, the professor does not commiserate: "You have been unlucky. Chance has not favored you." No, the student is admonished: "You have been sloppy. Go back and try again." Life, we have seen, is explained in chemical terms; so must its origin be.

For the reasons I have just summarized, I favor the view that life was

4. How Did Life Arise?

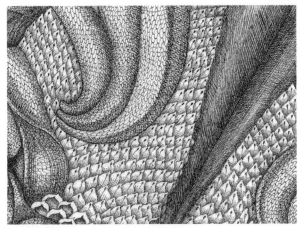

FROM WHAT
we have seen
in the pre-
ceding chapter, the most likely answer to the above question is: by a large
number of chemical steps that had a high probability of taking place
under the prevailing conditions. Alternative explanations, such as instant
creation or the intervention, at some stage, of a fantastic stroke of luck,
cannot be excluded as long as the postulated steps have not been identified; but they are heuristically sterile and unsupported by what is known
of the nature of life.

SIGNPOSTS ON THE WAY

The details of the life-generating pathway still elude us and may do so for
a long time. But they are not hidden in total darkness. First, we have a
pretty good idea of what the starting and ending points were. The former
consists almost certainly of the amino acids and other organic materials
that arise spontaneously in various parts of the cosmos. To believe otherwise would stretch the boundaries of likelihood excessively, considering
the close chemical kinships that exist between those substances and biological constituents and considering their apparent ubiquity. As to the
ending point, it is represented by the common ancestor of the whole
living world, most likely, as we have seen, a primitive bacterium already
endowed with all the basic properties that characterize present-day life.

We know the beginning and the end. But that is not all. We actually
know *one* way of getting from one to the other by *natural* means. It

consists of the universal mechanisms whereby life makes more life on Earth today. A number of investigators engaged in origin-of-life research believe this information to be irrelevant. Prebiotic chemistry, they feel, must have been very different from biochemistry. This is most likely true for the cosmic chemistry to which synthesis of the starting building blocks is attributed. But at some stage, the initial chemistry must perforce have given place to biochemistry. My reasons for assuming, against a widely held opinion, that this transition took place early, rather than late, will become clear as we progress in our analysis of the problem. In the meantime, let us start with something on which virtually everyone agrees.

THE ADVENT OF RNA REPRESENTS A WATERSHED IN THE ORIGIN OF LIFE

In the first chapters, attention was drawn several times to the central position of RNA in the blueprint of life. In all known living beings, genetic information flows from DNA to RNA to proteins. Striking in this sequence is the uncircumventable position of RNA, which is the obligatory intermediate in the expression of every bit of genetic information stored in DNA. This expression occurs invariably by transcription of the DNA text into the corresponding RNA, the DNA itself being essentially inert from the functional point of view. In a small number of instances, the transfer of information stops there. The RNA transcript plays a functional role by itself, as a ribozyme, or catalyst of a reaction. Most often, the RNA acts as a messenger. It instructs the synthesis of a protein, which, itself, by its structural qualities, or by its enzymatic properties, or by both, plays in the organism the role governed by the transcribed DNA segment.

It is striking and, no doubt, significant that the protein synthesis machinery actually contains RNA molecules as essential components. These are, in addition to messenger RNAs, the ribosomal RNAs, which are key catalytic constituents of the particles (ribosomes) on which proteins are assembled, and the transfer RNAs, those remarkable molecules that serve both to provide amino acids to ribosomes in a form suitable for the assembly of proteins and to read, by anticodon-codon interactions, the instructions borne by the messenger RNAs.

Compared with these crucially important functions, those fulfilled by DNA would seem to be rather minor, being restricted to the storing of information in a replicable (and transcribable) form.[1] In reality, this function could very well be carried out by RNA itself, which we have seen can be replicated, similarly to DNA and according to the same kind of complementarity relationships, by some viral enzymes. This does not mean that DNA is useless. Its dominant presence in all living beings is sufficient proof of its indispensability. But what is eloquently suggested by the facts is that *RNA preceded DNA* in the development of life and played for a while the role of replicable repository of genetic information carried out today by DNA.

It seems likely that *RNA preceded proteins* as well, considering the importance of the functions accomplished by RNAs in protein synthesis. Here, however, a clarification is in order. Proteins, as we have seen, are made from 20 different kinds of amino acids, which are the same in the whole living world. Now, many other amino acids exist, even in the products of cosmic chemistry and in those of the simulation experiments that made Stanley Miller famous. Some of these amino acids are found in biological substances other than proteins, sometimes even linked by peptide bonds of the kind that serve to join amino acids in proteins. There thus has happened, at some stage in the development of protein synthesis, a sort of selection that retained certain amino acids as building blocks for the RNA-dependent machinery and excluded others. We shall see later how this selection could be explained. Let us just, for the time being, remember that a distinction must be made between *peptides* and *proteins*. Peptides comprise all the substances, including proteins, consisting of amino acids joined together by peptide bonds. Proteins represent a subset of peptides, containing molecules of large size constructed exclusively with the 20 so-called proteinogenic amino acids, for which there are codons in the genetic dictionary. When proteins are said to have been preceded by RNA, it is that subset that is referred to, not the complete set of peptides. Indeed, it is very possible—I tend to say probable—that certain peptides may have preceded RNA, as will be seen later.

These considerations have led to the notion of an "RNA world," a term coined in 1986 by the American chemist Walter Gilbert, inventor of one of the first methods for sequencing DNA. According to Gilbert's definition, the RNA world represents a hypothetical stage in the development of life in which neither DNA nor proteins existed and RNA molecules alone carried out the functions of these two substances. They served as replicable support for genetic information and accomplished by their catalytic (ribozymatic) properties "all the chemical reactions necessary for the first cellular structures." This notion has met with enormous success and goes on inspiring numerous experimental attempts aimed at extending by engineering the catalytic capacities of RNAs, which, in nature, are largely restricted to protein synthesis and RNA processing.

We shall see that there are some difficulties with the RNA world as defined by Gilbert. But the foundations of this notion seem indubitable. There is every reason to believe that the emergence of RNA was a crucial step in the development of life, which preceded and most probably determined the appearance of DNA and of proteins. But, before RNA, there must have been something else that prepared and caused the advent of this key substance.

The Road to the RNA World

Incipient life, unless guided by a directing principle of the "intelligent design" kind, excluded *a priori* from our working hypothesis, did not have available the information we possess. It did not "know" it was going to invent RNA and, with it, a new language that would affect the whole history of our planet, perhaps even of the universe. It did no more than blindly follow a pathway imposed by the physical and chemical conditions that prevailed locally. It is not objectionable for us to call on our knowledge of the outcome of those events in our attempt to retrace their course, provided we keep clearly in mind that only efficient causes, not final causes, can have determined them. The problem, it must be acknowledged, is of daunting complexity. Without going into details of chemical structure, let it simply be said that the spontaneous genesis in some "primeval soup" of a molecular arrangement like RNA defies chemical common sense. Indeed, it has so far defied the ingenuity of chemists.

THE RNA ENIGMA

For several decades, some of the best chemists in the world have vigorously addressed the problem of the prebiotic synthesis of RNA. Until now, their efforts, however determined and imaginative their approaches, have not been encouraging. Experts are beginning to lose confidence in an undertaking aimed directly at RNA. They now toy with the idea that

RNA may have been preceded in its primordial functions by structurally analogous compounds likely to have arisen more easily.[2]

Some unconditional supporters of the original version of the RNA world take refuge in the notion of a "flick of chance." They imagine a few RNA molecules arising somewhere by an almost miraculous combination of circumstances. Such an event would, in their eyes, have been enough for the whole process to be launched, thanks to the ability of RNA to self-replicate and display catalysis. Such a view does not hold water. First, the very hypothesis of RNA arising by some chance event is chemically implausible. Moreover, having a little RNA obviously does not suffice for making more. The term "self-replication" is misleading in this respect, as it confounds two entities: information and synthesis. RNA provides only the former. For the latter, complex building blocks, energy, and strong chemical support are required. These conditions must have been satisfied already at the time RNA first appeared, since this substance could not have been replicated otherwise. They manifestly continued to prevail during all the time—at least centuries, if not millennia or more, as we shall see—when RNA dominated the scene. We are far from the fortuitously stabilized and amplified product of some random fluctuation.

If we follow this reasoning, we arrive at the conclusion that RNA arose in a chemical environment that was already of considerable complexity and included all the elements needed for this event and its perpetuation. It is interesting to recall in this connection the remarkable relationship, already mentioned previously,[3] that exists in today's living world between information and energy. At the heart of both we find ATP and its analogues, GTP, CTP, and UTP.

Indeed, in the synthesis of RNA, those four molecules provide the nucleotide units—AMP, GMP, CMP, and UMP—that make up the building blocks of any RNA molecule. In this reaction, triphosphates (NTPs) become monophosphates (NMPs), the two supernumerary phosphates being released as inorganic pyrophosphate (PP_i), while enough energy is made available to support the linking of the nucleotides to each other in the RNA chain.

On the other hand, we have seen that ATP is the universal conveyer of biological energy. What has been mentioned only in passing is that ATP is sometimes replaced in this function by one of its analogues. Thus, GTP fuels the mechanism whereby the messenger RNA tapes are moved

through the ribosomes. CTP provides energy for the formation of phospholipids, the main constituents of biological membranes (see Chapter
6), while UTP serves a similar function in the synthesis of a number of
complex substances formed from sugar molecules (polysaccharides). And,
as just mentioned, the four NTPs also provide the energy for the assembly of RNA (analogous reactions are involved in DNA synthesis).

There can be no doubt: *biological energy and information are intimately
linked* in today's living world. In all likelihood, this relationship goes back
to the very origin of the processes we are attempting to explain. Such
being the case, two possibilities may be considered, depending on
whether information is taken to have arisen from energy, or the opposite.
We shall ignore, for simplicity's sake, the third possibility attributing the
origin of both energy and information to a phenomenon without equivalent in present-day life. This question is rarely discussed. But it seems
to me that if one defends the notion of a primitive RNA, fruit of an
extraordinary combination of circumstances or of some unknown chemistry that remains to be discovered, the logical implication is to assume
that ATP and its analogues arose from RNA and, therefore, that information preceded energy (in its present form). Personally, I find this possibility highly unlikely. Given the need, underlined earlier, for a solid
chemical underpinning to support the RNA world during the whole of
its long evolution, it seems to me much more plausible to suppose that
ATP and its analogues belonged to this underpinning and, perhaps, already served in it as energy vehicles. Consequently, to resolve the RNA
enigma, we must go back to the primitive chemistry that functioned,
presumably with the help of ATP and its analogues, before RNA existed.
What must be searched for first is how some sort of primitive metabolism, a *protometabolism*, could have arisen spontaneously under prebiotic conditions.

TRACING PROTOMETABOLISM

A detailed examination of the chemical reactions that may have composed protometabolism is out of the question. Solid knowledge on this
subject is virtually nonexistent, anyway, and the speculations that stand
in lieu of it are almost as numerous and varied as the investigators interested in the problem. I shall content myself with a general remark. It
expresses a personal and far from widely accepted opinion, which, however, I will try to justify later: protometabolic pathways prefigured the

pathways of present-day metabolism. In other words, the signposts mentioned in the beginning of this chapter must be heeded right from the
start.

This affirmation, which I have called the *congruence principle*, implies
as an important corollary that present-day metabolism holds traces of the
primitive chemistry and could serve as a valuable source of inspiration in
the elaboration of theories and, especially, in the design of experiments.
Being, unfortunately, past my time for the latter, I must content myself
with the former.

The main lesson of metabolism was underlined in Chapter 1 (p. 19):
"virtually none [of the reactions of metabolism] would take place if the
participating substances were merely mixed together." It is for this reason
that most experts are skeptical of the congruence principle. In their opinion, prebiotic chemistry, not having available the catalysts of biochemistry, could not possibly reproduce the reactions of biochemistry. But one
may, instead, wonder whether appropriate catalysts could not have been
present in the cradle of life.

Needless to say, the search for possible prebiotic catalysts has always
been an important preoccupation of origin-of-life investigators. But their
search has, for obvious reasons, been largely restricted to the mineral
world; and it has not been entirely fruitless. Clays, in particular, have
proved capable of catalyzing the linkage of activated nucleotides into
small RNA-like associations, whereas certain iron-sulfur combinations
have been found to promote some reactions involving electron transfers.
However, nothing comparable to even a very primitive protometabolism
has ever been reproduced.

In nature, as we have seen, metabolic reactions are catalyzed mostly
by protein enzymes, often acting in conjunction with metals and with
organic coenzymes. Catalytic RNAs (ribozymes) are involved to a small
extent. In the original RNA-world view of Gilbert, ribozymes are taken
to do the entire job. It is, however, obvious that RNAs could not have
served as catalysts in a pre-RNA protometabolism. Furthermore, the catalytic properties so far observed with ribozymes are rather limited; they
do not show the diversity one would be entitled to expect for a metabolism-like system.[4] These facts have not, however, damped the ardor of
the more enthusiastic supporters of the original version of the RNA
world. The possibility that a wider gamut of catalytic RNAs may have
existed in prebiotic days has prompted a number of highly ingenious

efforts at extending the catalytic potentialities of RNA molecules by bioengineering techniques. These experiments have yielded fascinating results, but their relevance to the origin of life is questionable.

Strangely enough, proteins—or rather peptides, since true proteins must have come later (see p. 59)—have not, by far, enjoyed the same popularity as RNAs as potential prebiotic catalysts. This is surprising, considering the fact that amino acids may have been abundantly present in the prebiotic world, where they could have associated into peptides by relatively simple mechanisms.[5] In addition, peptides, being closely related to proteins, are most likely to include molecules with catalytic properties similar to those of protein enzymes.

On the basis of these considerations, I proposed, a number of years ago,[6] that the catalysts of protometabolism may have been peptides, or, rather, *multimers*, as I have called them to indicate that they could have contained substances other than amino acids but chemically close to them, for example, hydroxy acids. An objection to this hypothesis is that the postulated molecules would probably have been too small to display the required catalytic properties. But this objection is not necessarily valid since, as will be seen in the next chapter, the first protein enzymes were probably quite short, little more than about 20 amino acids long. This indicates that peptides of such short length, perhaps even shorter, may be endowed with catalytic activities, rudimentary to be sure, but sufficient to serve as primitive enzymes. Another objection is that a mixture containing all the required catalysts, assuming it had arisen by some chance circumstance, is not likely to have been faithfully reproduced for a long enough time without some replication mechanism. This objection, however, applies to any model of pre-RNA protometabolism, which would be subject to the same constraints. Environmental stability is a common condition of all models postulating a natural development of life.

The fact remains that the multimer hypothesis is no more than a conjecture and will remain so as long as it has not been subjected to experimental testing. This has become possible. Techniques now exist for the preparation of mixtures containing a large number of peptides of different structure. It would be possible to look for enzyme-like activities in such mixtures. This is what I would do if I were 20 years younger.

RNA COULD BE THE FIRST FRUIT OF NATURAL SELECTION

Leaving aside the question of mechanisms, let us return to the central notion, based on the congruence principle, of a metabolism-like protometabolism. The assumption is that ATP and other NTPs somehow arose—the details of possible reactions are beyond the scope of this book—as products of this protometabolism and became integral parts of it, possibly participating in reactions that prefigured their future bioenergetic role. It would not be surprising in such a context if some of the NTPs reacted together to make RNA-like associations.[7] This, it should be noted, would be a purely chemical reaction, explainable simply by the presence of a suitable catalyst. For the associations to be authentic RNAs, there would have to be intervention of a template molecule interacting with the catalyst so as to dictate, by base pairing, the choice of the reacting NTPs. UTP would be selected in front of A in the template, CTP in front of G, GTP in front of C, and ATP in front of U (see Chapter 2). Easy to imagine, you might say. But watch out! Here is where hindsight can be dangerously misleading.

Why just A and U, G and C? The possibility that chemical determinism happened to be such as to single out those two pairs of complementary bases smacks perilously of *pre*-determinism. Do we have to assume that "intelligent design" prepared the way to information transfer by guiding the atoms to combine in just the kind of molecules that allow pairing? Not necessarily. It seems much more likely, if, as would be expected, relatively unspecific chemistry was involved, that a whole array of kindred molecules[8] were produced besides the four canonical bases. Molecules of this kind exist today in living organisms.[9] Rather than endowing prebiotic chemistry with prophetic insight, it seems more probable that it indiscriminately made a variety of compounds of the same kind, including their NTP derivatives, and that, in turn, the RNA-like products of NTP combination included a "gemisch" of many different assemblages. If this is what happened, all we need is a couple of trivial assumptions, and the RNA "miracle" is explained.

Just imagine—surely a plausible possibility on a purely statistical basis—that a few molecules in the gemisch happened to contain, like authentic RNA, no other bases than A, G, C, and U. If such molecules could interact with the catalyst responsible for the assembly reaction in the manner postulated above, then complementary molecules likewise containing only the four canonical bases would be formed. These molecules, in turn, could induce the reproduction of the original molecules, and so on. Continuation of this phenomenon would progressively lead to the formation of an increasing number of complementary molecules of both kinds. What we have is selective *replication* and *amplification* of the rare true RNAs present in the mixture.

This mechanism thus accounts in one shot and without calling on any special intervention, whether of chance or of the deity, for the birth of RNA and for its first replication. As proposed, RNA no longer arises as the product of an almost miraculous event. It is formed by chemistry, as required. But it becomes dominant thanks to a new process, *molecular selection*, based itself on replicatability. This was a decisive turning point in the development of life. Until then, chemistry was solely in charge. To be sure, continuity was guaranteed by the strict determinism to which chemistry is subjected; but it was, for the same reason, exposed to the vagaries of the environment. With the advent of replication, the faithful reproduction of molecules became possible even under changing environmental conditions. The first seed of genetic continuity was planted.

But there is more. Primitive replications were no doubt very imprecise, continually producing imperfect replicas of the models. Among these faulty copies, there must have been some that, for various reasons, were more resistant to degradation than the originals or were replicated faster than them by the catalyst responsible for the synthesis of the first RNAs. In both cases, the molecules concerned tended to become more abundant than the others. As a consequence, the initial RNA mixture arising from the first products of prebiotic chemistry was to become progressively dominated by RNA molecules that combined stability and replicability in optimal fashion.

This is not just a theoretical vision. The molecular selection of RNA can actually be reproduced in the laboratory.[10] This feat was accomplished for the first time in the 1960s by an American biochemist, Sol Spiegelman, and has since been repeated under various conditions by a number of investigators, among them the German chemist Manfred Eigen, who has made a particularly detailed study of the phenomenon. These investigations have clearly established that the mechanism involved does, indeed, consist of a molecular selection entirely ruled by the combined criterion of stability-replicability of the molecules.

This mechanism, it should be emphasized, represents at the molecular level exactly that imagined by Darwin to account for biological evolution: diversification by modifications of the material responsible for hereditary continuity, natural selection of the modified forms most apt to survive and multiply under prevailing conditions, and amplification of those forms. But molecules, not organisms, are selected in this way, with RNA as first fruit of this fundamental mechanism.

The molecular selection of RNA taking place under the conditions of the prebiotic era must have led in the end to a dominant sequence that henceforth remained unchanged—the one combining stability and replicability optimally for those conditions—accompanied by a continually shifting cohort of sequences modified by replication accidents. Eigen has called such a mixture a "quasi-species." He has arrived, by investigations too specialized to be described here, at the conclusion that the dominant molecule in the quasi-species formed by the first RNAs, the "UrGen," or original gene, probably corresponded to the ancestor, as identified by molecular phylogeny analyses (see Chapter 7), of the whole family of transfer RNAs. It will be seen that this identity could be highly significant.

THE RNA WORLD

The hypothetical scenario just sketched out—or any other obeying the same criteria—shows how incipient life could have entered a phase that could rightly be called "RNA world," though not—at least, not yet—an RNA world supported by RNA catalysts, as proposed by Gilbert, which it obviously could not be at birth. RNA could not have served originally to make RNA. Whether it ever did cannot be excluded but is so far entirely unsupported by evidence. What seems highly probable, on the other hand, is that RNA served to make proteins. This will be the subject of the next chapter.

5. How Did Life Arise?

From RNA to Protein-DNA

A TTENTION HAS ALREADY BEEN DRAWN IN THE PRECEDING chapter to the fact that the protein synthesis machinery contains several essential parts of RNA nature. There is every reason to believe that proteins, as defined in the preceding chapter—polypeptides made from 20 specified kinds of amino acids—are an "invention" of RNA. Needless to say, the word "invention" is not intended here in its usual sense. What is meant is that the first true proteins arose from chemical interactions between RNA molecules and amino acids. Surprisingly, this central problem has so far hardly been tackled experimentally. What we have to guide our speculations is mostly our understanding of how the present-day machinery operates. Within the framework of the congruence principle, such a contribution can still be of great value.

How RNA Made the First Proteins

This process probably went through several stages. First, the chemical machinery must have appeared. Remember our governing principle: chemistry came first. Next came information, with the development of translation and the genetic code. Finally, from the early proteins must have arisen the first enzymes that initiated metabolism. I can, for obvious reasons, cover this complex topic only in a highly superficial manner.

THE CHEMICAL BIRTH OF PROTEINS

Any model for the development of protein synthesis must necessarily start with direct interactions between RNAs and amino acids.[1] One does not see how else RNA could ever have become involved in assembling amino acids. Most likely, certain RNA molecules became linked with certain amino acids. This, remember, is exactly what happens today between transfer RNAs and amino acids. Eigen's investigations, cited earlier, identifying the first RNAs as the ancestors of transfer RNAs, are particularly suggestive in this respect.[2]

It thus seems reasonable to suppose that the first RNAs appearing in the prebiotic milieu met in it an abundance of amino acids, together with conditions conducive to the sealing of these encounters by chemical bonds. Leaving aside the mechanisms that could have effected such linkings,[3] let us take a closer look at their specificity. Did the supposed associations take place randomly between any kinds of RNAs and amino acids? Or, on the contrary, was the process selective? The second possibility is by far the more attractive because it allows much to be explained.

Such a selectivity would, notably, provide an answer to a question many consider one of the great mysteries posed by the origin of life: why are proteins made from the 20 amino acid species that serve universally for their synthesis? It is not a question of relative abundance. Some amino acids present in large quantities, both in meteorites and in the products of abiotic syntheses, are not used for protein assembly. Others, though rare, are. In addition, the amino acids that participate in protein construction all belong, with one exception, to a class of molecules called chiral (from the Greek, *cheir*, hand). This term recalls the fact that the molecules can exist in two configurations that are to each other as the right hand is to the left hand, or, alternatively, as an object is to its image in a mirror. Proteins are made exclusively with one variety, designated L, as opposed to the other kind designated D. A likely possibility is that amino acids were selected for protein synthesis by virtue of their ability to interact with RNAs. This explanation provides a simple answer to the riddle of the proteinogenic amino acids, even though it may not entirely account for the chirality problem.[4]

Selection works both ways. If RNAs selected the amino acids used for protein synthesis, the amino acids must reciprocally have selected the RNAs serving for their transport. Chemical affinities are mutual. The selection process, however, could not have been symmetrical. The amino acids presumably were there from the start for the RNAs to choose from, products of cosmic chemistry and, eventually, of metabolism. The RNA molecules, on the other hand, had to arise by selective replication from a cohort of ever-changing molecules (see preceding chapter). This implies that the fact of being linked to an amino acid increased the stability, or replicability, or both, of the RNA molecules involved, with, as a consequence, their amplification by molecular selection. This hypothesis is compatible with what is known of the chemical aspects of such an association.[5] It would explain the molecular selection of RNAs capable of associating with amino acids and thereby driven eventually to become the transfer RNAs of living organisms.

Admittedly, this is all hypothetical. But the hypothesis rests on undeniable foundations and has the advantage of suggesting experimental approaches. Evidently, what needs to be studied is the manner in which amino acids (or their derivatives, such as thioesters) can interact with RNAs and, perhaps, influence their replication. Surprisingly, very few investigators seem to be interested in this problem, even though it could be investigated with current techniques for studying RNA evolution in the test tube.[6]

In order to give rise to proteins, the amino acid molecules attached to their transporting RNAs would have to become linked to each other by peptide bonds. As we saw in Chapter 2, this is accomplished in present-day living organisms by ribosomes in association with messenger RNAs. Ribosomes consist of approximately equal amounts of protein and RNA molecules, but there is good evidence indicating that the actual linking together of the amino acids is catalyzed by an RNA molecule, not by a protein. Considering, in addition, that no proteins could have been around when the events we are trying to reconstruct took place, it is tempting to assume that RNA molecules, presumably ancestral to today's messenger RNAs and ribosomal RNAs, made up the first protein-synthesizing machinery, perhaps in combination with certain multimers. How this could have happened will be examined later in this chapter.

According to the proposed model, RNA molecules ancestral to the three kinds of RNAs that take part in protein synthesis today were involved in this process right from the start. This hypothesis is consistent with the notion, suggested previously, that proteins were "invented" by RNA. But, as for the amino acid-binding RNAs, it raises the problem as to how the RNAs that participated in the early protein assembly machinery came to be selected. Being part of a complex scaffolding could conceivably stabilize the molecules and thereby contribute to their selection. Whether this would have sufficed is questionable. In any case, with the appearance of protein synthesis, incipient life reached a stage in which, for reasons that will become clear, the selection of RNAs could no longer be explained exclusively by their intrinsic molecular properties. Henceforth, *cells* became necessary for further progress.

INTERLUDE: INDISPENSABLE CELLS

From what we have seen so far, replicating RNA—or any other kind of replicating molecule, for that matter—is automatically and obligatorily subject to a selection process that favors, among the variants arising from replication errors and other accidents, those molecules that optimally combine stability and replicatability under prevailing conditions. In the proposed model, this process accounts for three successive selection events that, in turn, led to the first replicatable RNA molecules, thence to the "UrGen," the original genetic quasi-species held by Eigen to be ancestral to transfer RNAs, and, finally, to a set of RNA molecules capable of specifically binding amino acids.

In all those cases, the RNA molecules are assumed to be selected by virtue of their intrinsic properties. But there is a limit to such a mechanism. At some stage, selection of the RNA molecules must become based, not on what they *are*, but on what they *do*, that is, making proteins. This process requires an indirect mechanism, called hypercycle by Eigen, whereby the RNA molecules involved are selected by the protein molecules they help to make. The formation of physical hypercycle complexes could conceivably explain such effects, but not as a general mechanism. What is most likely is that the universal Darwinian mechanism of *cellular selection* took over at this stage. This requires the RNAs and their protein products to be segregated together in a large number of discrete units capable of competing with each other.

We shall see in the following chapter that origin-of-life specialists do not agree on the stage in the development of life at which the first cells appeared. According to some, it all started with cells. Others consider cell formation a late event. All, however, would agree that the beginning of protein synthesis represents an extreme limit beyond which life could not have proceeded further without cellularization.

The first cells were obviously much simpler than present-day cells. We shall call them "protocells." The minimum required to make a protocell capable of participating in a selection process is a set of replicatable RNAs engaged in the sort of amino acid assemblies we have just sketched, enclosed within an envelope together with all that is needed for their replication, including the required protometabolic support. In addition, these protocells must, like present-day cells, have been able to grow and to multiply by division, competing among each other for available resources. This implies the ability to extract from the environment the materials and energy they needed and to discharge waste products into it.

One readily imagines such a collection of protocells engaged in a Darwinian type of competition dependent on the proteins they are making. As a result of replication inaccuracies and other accidents, different protocells will perforce acquire different RNA variants. Suppose some of these RNAs cannot make proteins, whereas others can. To the extent that possession of proteins is favorable to protocellular reproduction, the protocells capable of making proteins will be advantaged in the competition. In turn, those among the protein-making protocells that make more proteins or make them faster will gain over the others. Such will especially be the case for the protocells that make *better* proteins. Thus, thanks to cellular selection, RNAs most useful by way of the proteins they are making will be favored.

You will notice that the preceding paragraph contains a shift from the quantitative to the qualitative. There is mention first of "more" proteins, then of "better" proteins. This transition reflects one of the great moments in the origin of life: the advent of translation and of the genetic code.

THE ORIGIN OF TRANSLATION AND OF THE GENETIC CODE
Much has been written, but few experiments have been done, on the origin of translation and the genetic code. I shall limit myself to two aspects of a general nature. First, there is the *structure* of the genetic code, which, it will be remembered, consists of a list of correspondences between amino acids and specific base triplets represented, in complementary form, by the codons of messenger RNAs and by the anticodons

of transfer RNAs. The simplest hypothesis to account for these correspondences is to assume that they reflect the primeval chemical affinities
that, as we have seen above, presumably ruled the mutual selection of
amino acids and of their transporting RNAs; in other words, to assume
that future transfer RNAs selected the amino acids used for protein synthesis by way of their anticodons. This attractive hypothesis has few
defenders today. The required chemical complementarities between anticodons and amino acids do not seem to exist.[7]

Another hypothesis, likewise out of favor today, is that proposed by
Crick under the term "frozen accident." According to this hypothesis,
the correspondences between amino acids and anticodons, and thus codons, arose by chance, to be subsequently sealed by usage.

The theory considered most likely today supposes a historical, coevolutionary process in which the anticodons and the corresponding
amino acids were progressively recruited together under the control of
natural selection. Several arguments support this hypothesis. The most
convincing lies in the structure of the code, which, far from being random, happens to be such as to minimize the deleterious consequences of
mutations. Indeed, in many instances, replacement of one base by another
in a messenger RNA codon leaves the nature of the inserted amino acid
unchanged—remember, the genetic code contains many synonyms—or
causes it to be replaced by an amino acid sufficiently similar to it in its
physical properties for the modified protein to remain functional. It has
been shown by theoretical model analyses that the present code is close
to optimal in this respect.[8] Unless intelligent design is invoked, such
optimization can be explained only by a competition among many different codes, with the verdict being rendered by natural selection. The
genetic code that has emerged is essentially universal, the only rare exceptions being due to minor changes that occurred late in evolution.[9]

The second point I wish briefly to address is the origin of translation
itself, which may not be as mysterious as it appears at first sight. In
descriptions of the function of messenger RNAs, it is customary to emphasize the "messenger" aspect of the molecules, that is, their *informational* role, neglecting the fact that they also play an important *confor-
mational* role in the physical organization of the ribosome system. By
pairing with the anticodons of the transfer RNAs, the codons of the
messenger RNAs not only *select* the appropriate amino acids; they also
help to *immobilize* the transfer RNAs on the surface of the ribosome in

an orientation conducive to the joining together of the carried amino acids by the catalytic ribosomal RNA component. It is quite possible, perhaps even likely, that, in the primeval RNA scaffolding believed to be responsible for the synthesis of the first proteins, the ancestral precursors of messenger and transfer RNAs already interacted by something akin to codon-anticodon pairing. This is a plausible hypothesis, considering that base pairing is the main mechanism whereby RNA molecules interact. Thus, if proteins were first made by a machinery consisting of interacting RNA molecules, base pairing was almost certainly involved.

If the proposed hypothesis is correct, translation by codon-anticodon pairing took place in some primitive fashion from the very beginning of protein synthesis. The sequence of the resulting protein, however, depended on whatever "code" existed at the time, that is, on which amino acids were borne by carrier RNAs with different anticodons. From what we have seen, this original code probably had little to do with the present code, which shows clear evidence of being the outcome of a long selective optimization process. It is interesting to consider the implications of such a mechanism.

Let us start with a first code, whatever it was. The protocells obeying this code will be subject to two kinds of mutations, depending on whether a message or the code is affected. The first kind of mutations will cause changes in only one protein, the one encoded by the altered message and produced by translation from it in accordance with the existing code. Competition among protocells subject to such mutations will lead to the selection of those that make the most useful proteins by way of the particular coding system in use. The second kind of mutations will have much more sweeping consequences; they will simultaneously affect all the proteins in which some amino acids have been replaced by others as a result of the change in the code. To illustrate the difference, read this paragraph replacing "c" by "m" in only the word "code" (message mutation) or everywhere in the paragraph (code mutation). In the first case, the paragraph remains readable and even, stretching definitions a little, understandable. In the second case, it is gibberish. Thus, one would expect most code mutations to be lethal. But it could happen, on rare occasions, that a change in coding produces a new set of functional proteins, allowing the new code to survive and to compete with the existing one. Such an event is likely to occur only in primitive systems containing few proteins and admitting a certain leeway in the precision of translation. With a large number of proteins and a strict translation system, changes in coding are almost bound to be lethal. It is almost certain, in such a case, that at least one indispensable protein will be rendered inactive. It is this kind of consideration that has inspired the frozen accident theory. Once a code is in place in a sufficiently complex system, it can no longer be changed.[10]

The fact that the present genetic code appears to be optimal suggests strongly that the early history of protein synthesis depended largely on code mutations,[11] that is, mutations in the amino acid-transporting RNAs.[12] The major diversification of proteins and the kind of Darwinian competition between genetic messages that is universal today can have flourished only after the final code had been adopted.

THE BIRTH OF ENZYMES

In the proposed model, the evolution of early protein synthesis is pictured as conditioned, in turn but with considerable overlap, by three different kinds of mutations. These affected, first, the efficiency of the RNA machinery, then, the resilience of the coding system, and, finally, the quality of the synthesized proteins. In the third kind, which eventually came to dominate the scene, the main criterion of protocell selection was the usefulness of the proteins arising by translation of mutated RNA messages, that is, their ability to favor protocell growth and proliferation. This, presumably, is how the first protein enzymes appeared. To be sure, these early enzymes were a far cry from present-day enzymes. They were probably little more than about 20 amino acids long, as we shall see later, and their catalytic activities must have been very rudimentary. Though crude, these activities must nevertheless have been sufficiently useful to favor the selection of the protocellular owners of the mutated RNAs. Let us look a little more closely at how this could have happened.

It all has to start with a mutation of an RNA message—due, for example, to a replication error—leading to the production of a modified protein that happens to possess a certain catalytic activity, a primitive enzyme. If the activity of this enzyme proves useful to the protocell in which the mutation occurred, this protocell will multiply faster than the others and its progeny will eventually become dominant. Let the same phenomenon be reproduced for another enzyme, and selection will now bring out protocells possessing the two enzymes. Step by step, repetition of the same mechanism will finally lead to a population of protocells fitted with a full set of enzymes capable of satisfying all their needs. These protocells will be freed at last from their dependence on the primitive chemistry that supported the RNA world during its evolution. Proto-metabolism has given place to metabolism.

It is difficult to account in any other way for the emergence of present-day metabolism catalyzed by protein enzymes. A crucial element of the envisaged scenario is the need for the retained enzymes to be *useful*, without which there could be no selection. In order to be useful, an enzyme must necessarily have available in its environment one or more substances—the technical term is "substrates"—on which to act. Without substrate, even the most sophisticated catalyst is valueless. The enzyme must also have an outlet for the products it forms, if it is not to run into a chemical dead end. These substrates and outlets must have been pro-vided by the primitive protometabolism that supported the protocells at the time. Or, put differently and more pertinently, only those enzymes that found substrates and outlets in the existing protometabolism could have been retained by selection. This protometabolism, therefore, acted as a screen for the selection of the first enzymes, which must, by necessity, have fitted within the existing chemistry. This, in my opinion, is a strong reason for believing, as I have stated before in this chapter, that proto-metabolism and metabolism were *congruent*, that is, followed *similar pathways*.

This opinion is not shared by many scientists engaged in prebiotic chemistry research, most of whom tend to follow the avenues of organic chemistry more often than those of biochemistry. The attitude of these experts is understandable, given the strict dependence of biochemistry on enzymes that could not have been present in the prebiotic milieu. Con-sidered from the point of view of organic chemists, the reactions that take place in living cells are definitely not of the kind such chemists would expect. On the other hand, the congruence argument cannot be ignored. As mentioned above, the prebiotic participation of catalytic mul-timers prefiguring present-day enzymes could possibly account for the metabolism-like protometabolism required by this argument.

What about the role, in the proposed model, of catalytic RNAs—ribozymes—to which the widely publicized image of the RNA world attributes such fundamental importance? Such a role is evident in protein synthesis, which clearly stands out as having almost certainly been orig-inally carried out by RNA molecules (with the help of multimers?). It will be seen later that catalytic RNAs probably played an important role also in the lengthening processes that have led to present-day genes. Here, again, revealing traces of such a function are found in present-day living beings. Have ribozymes, in addition, carried out, as many believe, a number of catalytic functions fulfilled today by protein enzymes? It is not impossible, but little in present-day living organisms supports this idea.[13] It is worth recalling that the first RNAs were the products of a complex chemical network in which, by definition, ribozymes had no part. In principle, therefore, this network, if sufficiently stable, could by itself have supported the entire evolution of primitive life toward the metabolic autonomy ensured by protein enzymes. It is, of course, possible, that this network came to be enriched by ribozymes in the course of its evolution.

The Growth of Proteins

At the stage we have reached in our attempt at historical reconstruction, emerging life started to resemble present-day life, with, however, two important differences. Genes were most likely made of RNA, not of DNA; and they must have been much shorter than today's genes. How did the long DNA genes that exist today and, as we have seen, must have existed already in the common ancestral form from which all known living beings are derived, arise? The present structure of genes offers some clues to this question.

THE FIRST GENES WERE VERY SHORT

According to Manfred Eigen, already cited several times, the first RNA genes were probably no more than some 75 bases long, the size range of transfer RNAs, which Eigen has tentatively identified as the direct evolutionary descendants of the primeval gene (see Chapter 4). This estimate, which is to be compared to the many hundreds of bases that compose today's genes, is consistent with theoretical calculations, also by Eigen, showing that the length of a replicatable molecule is limited by the accuracy of the replication system.[14] According to Eigen's theory, a length of 75 bases would correspond to a maximum replication error rate of 1.33 percent, not an unlikely value for the very crude system that must have operated in those early times.

There is an interesting corollary to these evaluations. If the maximum length of the first RNA genes was 75 bases and if allowance is made for a few untranslated bases at both ends of the molecule, the first protein enzymes produced by translation of those genes could have been little more than about 20 (one amino acid per base triplet) amino acids long. This means, contrary to what is sometimes stated, that peptides of such short size can display catalytic activities. These, no doubt, were much cruder than those of present-day enzymes but, as we have seen, must have sufficed to ensure all the needs of incipient metabolism. This fact lends credence to the proposal, made in Chapter 4, that small peptides— or closely related multimers—may have served as catalysts in protometabolism.

GENES GREW BY MODULAR COMBINATION

How did the primeval genes grow to tens of times their original length? If, as just seen, the accuracy of replication limits the size of the genes, their lengthening necessarily had to proceed by way of the development of more precise replication systems. This phase of evolution thus appears as contingent on mutations leading to replication systems of increasing accuracy. One may thus imagine a series of steps, characterized each by the appearance of a more reliable replication enzyme, allowing a corresponding lengthening of the genes and, therefore, of their protein products. To the extent that longer proteins can make better enzymes, each stage will thus lead to a general improvement in the efficiency of the protocells' enzymes.

It is very probable that this lengthening did not take place gradually, base per base, or amino acid per amino acid, but, rather, in modular fashion, by the combination of entire segments of RNA (or, perhaps, of DNA, see below) and thus, for proteins, by the combination of entire peptide blocks. Undeniable traces of such a process are found in the structures of present-day proteins, which are manifestly made of modules, or "motifs," of which many participate, diversely associated, in the construction of a number of different proteins. It is possible that the immense variety of present-day proteins is the outcome, further diversified by evolution, of the combination of only a few thousand distinct modules. An important aspect of this genetic combinatorial game is that it provides an answer to the "sequence paradox" already evoked in preceding chapters, namely the fact that life occupies a negligible fraction of the space of possible sequences, a place that, according to the defenders of intelligent design, it could not have reached without guidance. Consider the following elementary calculation.

Suppose that the first proteins had a length on the order of 20 amino acids and, which is far from certain, that they were already constructed with the 20 known proteinogenic amino acids. This amounts to a total of 20^{20}, or 10^{26}, different possible arrangements. The figure is high, but not inordinate. Note that if, as many believe, life started with a smaller number of amino acids, the estimate would be much smaller. But let us stick to our first figure: 10^{26}, or 100 million billion billion different possible proteins 20 amino acids long. It is an enormous number. Yet, if protocells had been the size of present-day bacteria—they could very well have been smaller—that number of protocells could have fitted, with 99.9 percent of the volume to spare, in a moderate-size lake measuring, for example, 20×50 kilometers in surface and 100 meters in depth. Reflect, in addition, that millions of successive generations of protocells may have succeeded each other during the time it took for incipient life to acquire its first full set of protein enzymes and one arrives at the conclusion that this outcome may have been the fruit of an essentially *complete exploration* of the available sequence space, reduced in the end to a few hundred by natural selection.

The next step, according to the proposed hypothesis, involves an essentially random combination of existing sequences, with, again, natural selection deciding. Admitting, for the sake of simplicity, the presence, at the start of this phase, of 1,000 different proteins of 20 amino acids each, we find that their combination two by two may yield a maximum of one million ($1,000 \times 1,000$) different proteins of 40 amino acids. Exhaustive exploration of this space is obviously possible. After adequate reduction of the number of sequences by natural selection, we once again reach a figure allowing the complete exploration of the subsequent space—comprising, for example, sequences of 60 or 80 amino acids—which, in turn, will be reduced to a manageable size by natural selection. Thus, by the repetition of the same process at an increasing level of complexity, life could have reached the infinitesimal place it occupies in an immeasurably immense sequence space by a course that involved, at each stage, the faculty of testing and submitting to natural selection virtually all the molecular sequences that were possible at that stage.

Needless to say, reality must have been considerably more complex than the highly schematic succession of events I have considered. But the principle is clear. The combinatorial mechanism of gene lengthening provides a solution to the sequence paradox. In lieu of intelligent design, natural selection has served as a guide among what, at any given time, was an essentially complete choice of available possibilities.

There can be no question of examining here the mechanisms that could have mediated this combinatorial game. Let me simply point out that the present-day living world holds numerous examples of RNA processing.[15] Remarkably, this is the second area, besides protein synthesis, in which ribozymes play a major role. It is tempting to assume that these ribozymes may be descendants of catalytic RNAs that participated in the lengthening of the first genes. This is one reason for supposing that this process involved RNA segments. But a process involving DNA segments cannot be excluded.

The Advent of DNA

The stage in the origin of life at which DNA appeared is not known. All that can be said is that DNA almost certainly came after RNA. We have seen that numerous arguments support this view. Also highly probable is the notion that DNA was derived from RNA. The two kinds of molecules are so similar that any other hypothesis hardly seems conceivable. Indeed, it sufficed, in order to pass from RNA to DNA, that two RNA constituents be replaced by two close relatives, ribose by deoxyribose, and uracil (U) by thymine (T). As to the information, one readily imagines that it may have been transferred from RNA to DNA by reverse transcription, as happens in cells infected with retroviruses, such as the AIDS virus. Two additional reactions were required for the DNA to become operational, replication of the DNA and its transcription to RNA. The reactions involved in all three processes resemble greatly the primary reaction of RNA replication. In all four cases, there is assembly of a polynucleotide chain (RNA or DNA) on a template made itself of either RNA or DNA, by a mechanism dependent on base pairing. The primitive enzyme that affected RNA replication could have served initially to catalyze these other reactions following some minor changes in its structure or, perhaps, without any such changes at all.

So much for mechanisms. What about the advantages that could explain the selection of the mechanisms? The most evident of these advantages is the division of the functions previously fulfilled only by the RNAs. Information storage and replication become the prerogative of DNA, while the utilization of this information for protein synthesis and other functions remains the province of RNA. Thanks to this division, all the genes could be regrouped, most often in single copies, in one long molecule, the first chromosome. And the replication of these genes could be carried out in synchronous fashion, coordinated with cell division, leaving to RNAs the faculty of performing their functions without impediment, such as, for instance, the formation of poorly reactive double-helical structures arising from RNA replication.

Another advantage of DNA is that it allows a selective expression of individual genes by way of transcription. This control takes place today by means of regulatory sequences interposed between the genes in the chromosomes. These sequences are acted upon by proteins, called *transcription factors*, that either stimulate or repress the transcription of the genes commanded by the sequences. As a result of these influences, individual RNAs and, therefore, proteins may be either produced or not, or produced in greater or lesser amounts, according to relationships that natural selection has adapted to the requirements of the cells. It may thus happen that certain enzymes are manufactured only when their need becomes manifest (by way of chemical circuits, needless to say). This kind of regulation, which is already important in bacteria, has become essential, as will be shown in Chapter 11, for the genesis of cells performing different functions in eukaryotes. Transcription, it should be noted further, most often involves only one of the two DNA strands. It thus yields single RNA strands that do not run the risk of being smothered into double helices by complementary strands.[16]

These advantages, to which one should add many others to be complete, suffice to demonstrate the irreplaceable character of DNA in all extant living beings. The benefits are so important that one would be tempted to suppose that DNA appeared very soon after RNA, especially as its advent does not seem to have required major chemical innovations. One may wonder, in this respect, whether the gene-lengthening process mentioned above could have taken place with DNA or had to occur with RNA. The first hypothesis is plausible, as many combinatory processes involving pieces of DNA are known. On the other hand, the importance and nature of the modifications undergone today by RNAs, in particular splicing, and the involvement of ribozymes in some of these processes, argue in favor of the second hypothesis, as we have seen. The uncertainty remains.

How Long Did It Take?

It was believed at one time that life may have required as long as several hundred million years to arise. Remember, it was this belief that led Crick to propose his "directed panspermia" theory. It is because he felt that there was not enough time on the prebiotic Earth for life to develop locally that he suggested an extraterrestrial origin (see Chapter 3). This argument is no longer considered valid by most workers in the field. In fact, the chemical nature of the processes involved makes it imperative for them to have been relatively fast. Otherwise, the many fragile intermediates that must have participated in the process could not possibly have reached levels compatible with their further utilization.

To illustrate this fact, let us consider the conversion of sugar to alcohol (and carbon dioxide). In nature, this process takes place by ten consecutive steps. Put in highly schematic terms, substance A (sugar) is converted into B, which becomes C, and so on, until the final product K (alcohol) appears. Suppose we add sugar (A) to an appropriate mixture of catalysts—yeast juice, for example, as in Büchner's celebrated experiment (see Chapter 3)—and follow the course of the process. We shall witness the successive rise of the levels of B, C, D, E, F, G, H, I, J, ending with the appearance of alcohol (K). Eventually, the system will adopt a state of dynamic equilibrium, or steady state—its cruising speed, so to speak—with the levels of all intermediates remaining constant or falling very slowly, while sugar progressively disappears and alcohol appears.

With concentrated yeast juice, which contains all the necessary enzymes in large quantities, this state could be reached in a few minutes. But if the amount of yeast juice added is decreased, the time needed to arrive at a steady state will increase, until the degree of dilution of the enzymes becomes such that the transformation of sugar into alcohol ceases to take place at a detectable rate. A given reaction of the chain—the conversion of F to G, for example—has become so slow that the intermediate G, deviated by side reactions (in particular spontaneous degradation), never reaches a sufficient concentration to allow its subsequent transformation into H to occur at an appreciable rate. Transposed to the prebiotic era, this example supports the statement that, in the unfolding of the chemical reactions that first led to life, each step must have been fast enough for the next step to be possible. From the moment the rate of formation of a fragile intermediate becomes too slow in relation to its rate of degradation, the process, perforce, grinds to a halt.

How fast is fast enough? How slow is too slow? It is practically impossible to answer this question without some fairly accurate model of the phenomena involved. All that can be said is that, if the pathways followed by prebiotic chemistry in some way resembled present-day metabolic pathways—the contrary, we have seen, is unlikely—the time taken by nascent life to move from the basic building blocks produced by cosmic chemistry to the first RNA molecules cannot possibly have numbered in millions of years, perhaps not even in millennia or centuries. A number of metabolic intermediates, such as ATP and the other RNA precursors, are much too fragile to participate in such slow processes. But is it a question of hours, days, months, or years? Without knowledge of how things happened, conjecture is fatuous. In any case, what counts most is not the time taken for the first RNA molecules to arise but the time during which the primitive chemistry had to support emerging life throughout the development of replication, protein synthesis, translation, the genetic code, and, finally, the first enzymes. Only at the end of this long road did nascent life reach a sufficient degree of metabolic autonomy to cease being dependent on the primitive chemistry. Let us just consider the last phase, in which enzymes were acquired.

The minimum number of enzymes needed to allow autonomous life of the present-day kind is estimated to be about 300.[17] Even if the protocells could do with a smaller number of enzymes of broader specificity, it is difficult to imagine autonomy with less than, say, 100 different enzymes. This means that, just to get out of the RNA world and acquire a first set of rudimentary enzymes, the protocells had to go through at least 100 selection rounds, initiated each time by a single mutant protocell. In each round, this mother protocell must have generated, by successive divisions, a population endowed with the new property, a population that becomes dominant, by selection, because of this new property. Another beneficial mutation occurring in this population then sets forth a new selective episode of the same kind, with the same sequence of events being repeated at least 100 times. With modern bacteria and optimal selection conditions, a history involving 100 successive selection events would probably demand a minimum of several months. This is to say that, with primitive protocells and environmental conditions no doubt far from optimal, acquisition of the first elements of metabolic autonomy may have required a considerable amount of time, to be measured in centuries, if not millennia, depending on the efficiency of the primitive chemistry that supported the protocells.

But this is not the end. For the protocells to evolve into something resembling bacteria, much more had to happen. The primitive genes had to grow to their present length by a process that, as we have seen, probably depended on a number of successive cycles of combination followed by selection. In addition, the genetic information had to be transferred from RNA to DNA, and the machineries required by this transfer had to be set into place. Altogether several additional millennia may well have been needed. This estimate should be tempered by the fact that the processes under consideration probably were continually accelerating, their main driving force being the gain in metabolic efficiency.

In conclusion, and granting the extremely rough nature of such evaluations, the complete pathway from building blocks to the first organisms possessing the basic properties of present-day life may have taken a time—probably to be counted in millennia or, possibly, in tens of millennia—very much shorter than the hundreds of millions of years proposed by earlier estimates. In fact, as already pointed out in Chapter 3, with a window of some 100 million years between the time Earth became physically capable of harboring life and the time the common ancestor of all life emerged, there may have been plenty of opportunities for living forms to arise—and disappear—at various times and in various places, before life finally took root.

Even so, the estimated order of magnitude represents a considerable amount of time, totally incompatible with the frequently presented image of the RNA world as a fleeting and precarious stage, rendered possible by exceptional conditions and rapidly leading to a stable situation secured by genetic continuity. Protometabolism must have rested on a robust chemistry, solidly supported by the prevailing physical-chemical conditions.

6. How Did Life Arise?

THE BIRTH OF CELLS

THE CELL IS THE unit of life. We have seen in the preceding chapter why cellularization was absolutely necessary for life to proceed beyond a certain level of chemical development. But we have not considered the possible nature of the mechanisms involved in this key step, nor the stage in the development of life at which it took place. The best way to approach these questions is to examine first the main requirements that need to be met in order for cells to exist.

THE MAKINGS OF A CELL

Cellular life depends on a number of fundamental properties that must have been achieved, albeit in primitive fashion, already in the first protocells. What is needed, in the first place, is a boundary, obligatorily endowed with the ability to exchange matter, energy, and information with the outside. In addition, cells must be capable of growing within the confines of their boundary and of multiplying by division.

THE HALLMARK OF A CELL IS A SURROUNDING MEMBRANE

The interior of cells is semifluid; it runs out readily as soon as the surrounding "skin," or membrane, is torn. This membrane forms a closed, saclike structure that envelops and keeps together the cell contents. Thus, the existence of a cell is contingent on the existence of a closed peripheral envelope. Everyday experience gives us a hint as to how this can occur. Whenever we use soap, we witness the creation of such structures, in the form of foam or bubbles. Physically, biological membranes are not very different from soap bubbles. In both cases, we are dealing with exceedingly thin, highly flexible, self-sealing films, only a few millionths of a millimeter in thickness. In both cases, the films consist at the molecular level of two opposed layers made of long, stick-like molecules that are closely packed parallel to each other and perpendicular to the plane of the layer, like bristles on a flat, doormat-like surface. The two layers always oppose each other by the same face, which allows two arrangements, bristles to bristles, or back to back, in our mat analogy.

The secret of such arrangements lies in the anatomy of the constituent molecules, which consist of a long, fatty tail made only of carbon and hydrogen, such as is found in petroleum hydrocarbons, attached to an electrically charged head. The tails of the molecules have, like hydrocarbons (and biological oils and fats), a strong tendency to avoid water and keep together. The heads, on the contrary, have a high affinity for water, which is a polar molecule to which they bind by electrostatic attractions. When such substances are agitated with water, they spontaneously organize so as to satisfy both kinds of preferences.[1] The tails stick to each other, forming the bristles of the mat, while the heads remain in contact with water, joining to form the back of the mat.

In soap bubbles, two layers of such molecules face each other by their heads, held together by a film of water, and the tails are in contact with air inside and outside the bubble. The reverse arrangement is found in biological membranes. In it, the two layers face each other by the tails of the molecules, which are intermingled to form a fatty, water-impermeable film. The heads of the molecules line the outer faces of the film, in contact with the two watery milieus present inside and outside the sac. Such structures are self-sealing, like soap bubbles. They always organize into closed sacs.

The membranes that surround living cells (and partition many into distinct compartments) are all made of such double layers, or *bilayers*. Their constituent molecules are known as *phospholipids*. As their name indicates, these substances belong to the general group of lipids, or fatty substances, from which they derive their fatty tails. In addition, they contain phosphate and other electrically charged groups, which make up their water-loving heads.

CELL MEMBRANES MEDIATE EXCHANGES WITH THE OUTSIDE

In themselves, phospholipid bilayers are little more than inert barriers that effectively separate from the outside the inside of the enclosures they delimit. They stringently curtail the exchanges of materials between the two. In cell membranes, these exchanges are ensured by proteins[2] inserted into the bilayer, which serve as entry and exit ports that permit, among other substances, nutrients to get in and waste products to get out. These ports often consist of several different proteins, organized so as to allow only certain specific substances to pass through. Some are fitted with energy-transducing systems, supported by ATP or by electric disparities, that can force given molecules to move against the direction of their normal flow. As is well known, molecules spontaneously diffuse from higher to lower concentrations, tending toward uniformity, the state of highest entropy. This tendency is countered, with the expenditure of energy, by what are known as *active transport systems*, or *pumps*.[3] Cells equipped with such systems can draw in rare materials from the environment or, conversely, drive back into the environment materials already present in it in high amounts.

The outer membranes of cells are also fitted with systems, called *receptors*, that allow the cells to respond to external chemical signals. Receptors usually consist of protein molecules inserted into the lipid bilayer and containing outward oriented sites capable of specifically binding certain substances, such as nutrients, hormones, or drugs, present outside the cells. As a result of this association—incidentally, a typical case of the mortise-and-tenon kind of molecular complementarity—the receptor molecules undergo a change in conformation, which, in turn, affects some system inside the cell. Receptors thus make it possible for outside substances to influence cells without entering them; they play a very important role, especially in multicellular organisms, where they allow different cells to communicate with each other by a highly intricate chemical language. Much of pharmacology depends on substances capable of interacting with certain surface receptors and thereby inducing some specific effects.

EXTERNAL WALLS OFTEN PROTECT CELL MEMBRANES

Most bacteria are surrounded by a rigid structure known as the *cell wall*. This structure, which is external to the cell membrane proper, is assembled from materials the cells secrete around themselves. In some instances, the wall is little more than an inert, porous shell serving mostly a protective function. In others, the wall's fabric is more complex, often lined inside by a membrane-like skin; and the space between wall and membrane, called periplasmic space, is the site of various chemical events, in particular the digestion of complex substances.

In the vast group of eubacteria, or Bacteria (see Chapter 8), this wall is made of a substance called murein, a gigantic mesh of interlinked sugar and amino acid molecules that completely surrounds the cell within a netlike structure. Interestingly, murein, in contrast to proteins, contains amino acids of D, as well as L, chirality (see Chapter 5). In the other main group of bacteria, archaebacteria or Archaea, the wall is made of a similar substance, pseudomurein, which contains only L-amino acids. Note that the wall is not an essential component of bacterial life. Naturally wall-less forms exist, sometimes in what could be taken as hostile environments, very hot ones, for example.

In eukaryotic life, walls are often replaced by more flexible outer coverings that permit surface deformations, such as are involved in movement or phagocytosis (see Chapter 10). Only the cells of plants and fungi, which are immobile, are surrounded by rigid casings, usually made of carbohydrate polymers, such as cellulose (plants) or chitin (fungi). Hooke's original "cells" (see Chapter 1) were no more than the remnants of the joined casings that were inhabited by cells in the living bark.

THE "INNARDS" OF CELLS

There is an enormous difference between the inside of eukaryotic cells and that of bacteria. As we shall see in Chapter 9, eukaryotic cells harbor a characteristic nucleus and a variety of membrane-bounded parts. Bacterial cells are much simpler. Many contain none of the structures seen in eukaryotic cells and offer better examples of the minimum needed to make a living cell. The most primitive such cells belong to the order Mycoplasmatales, often used nowadays in experimental attempts at identifying the minimum number of genes compatible with independent life. It is from such cells, deprived of various genes by experimental manipulations, that the figure of about 300 essential genes, mentioned in the preceding chapter, has been obtained.[4]

In bacterial cells, the genes are linked together in a single, often circular structure, or *chromosome*, anchored to the cell membrane. Essential components include all the enzymes needed for replication and translation of the genes, as well as the ribosomes and transfer RNAs required for the translation of genetic messages into proteins. In addition, the cells contain a core of metabolic enzymes, allowing the use of environmental materials, whether mineral (autotrophs) or organic (heterotrophs), for the retrieval and use of energy and for the synthesis of all cellular constituents.

THE GROWTH AND MULTIPLICATION OF CELLS

In a way, division is an almost obligatory correlate of growth. Consider a spherical cell. As it grows, its volume increases by the third power of its radius, its surface area only by the second power of the radius. A stage must necessarily be reached where the surface area becomes insufficient to support the exchanges with the environment needed for further growth. There are two possible solutions to this problem. The cell shape may change, allowing a greater surface-to-volume ratio. Or the growing cell divides into two smaller, spherical cells no longer subject to the limitations of the parent cell. The first solution is often observed, especially in eukaryotic cells. The second solution is universal.

In eukaryotic cells, cell division depends on highly intricate mechanisms, which will be considered in Chapter 9. The mechanisms are simpler in bacteria but, nevertheless, involve coordinated events directly linked to DNA replication. Once the chromosome has been duplicated, a set of processes are triggered, leading to progressive stricture that eventually divides the cell into two parts, each containing a chromosome and a full complement of enzymes and machineries needed for viability. In the course of this process, cell membrane and cell wall reorganize to form closed boundaries surrounding the two parts.

The Making of the First Cells

With the information summarized above, we may now examine how the external envelope of the first cells could have formed and at what stage this confining event took place.

THE FIRST CELL BOUNDARIES

Artificial phospholipid vesicles are readily made simply by subjecting a mixture of phospholipids and water to vigorous agitation, by means of ultrasounds, for example. Structures of this kind are made industrially on a large scale and marketed, under the name of *liposomes*, to serve as vehicles for cosmetics and drugs. It is thus easily imagined that the first protocells arose in a similar manner on prebiotic Earth, maybe not from phospholipids, which are fairly complex substances, but from simpler materials with similar physical properties. This theory, which is supported by the detection of traces of such materials in meteorites, is advocated by a number of scientists.

The theory is attractive but meets with a serious difficulty. Liposomes are very impermeable. They let through only molecules that are sufficiently fat-loving to pass through the thin, oily film of the lipid bilayer. One hardly sees how the RNA world could have developed and been maintained within such a hermetic enclosure, whatever the chemical processes involved and the stage at which cellularization took place. Substances had to enter the protocells to feed the metabolic and synthetic reactions that took place inside them, and the waste products of these reactions had to get out. Few of the molecules known to participate in these processes today have physical properties allowing them to readily traverse lipid bilayers. That things might have been different for the primitive processes would be surprising.

A possibility, evoked by Blobel, the German-American biologist already mentioned in Chapter 2, is that the vesicles were born empty and first served as support for externally attached prebiotic systems. In the course of their evolution, the flat vesicles would have progressively folded around the attached systems and ended up surrounding them with a double-membranous envelope, which would close only after acquiring a minimum of transport systems. In apparent agreement with this proposal, a number of bacteria are surrounded by two membranes, of which the outer one lines the inner face of the cell wall.[5] It is not known, however, whether this trait is an ancient heirloom.

Another possibility is that the first envelopes were not made of lipid bilayers but were porous structures possessing openings that let through foodstuffs and waste products while opposing the passage of the large molecules, RNAs and proteins, that had to remain inside. Those openings would have become progressively plugged with lipids as exchange systems were put into place. Some structure ancestral to the cell wall could have played this role.

There is no indication that cell membranes are derived from cell walls. But the structure of cell walls suggests a possible model for an ancestral porous envelope made of amino acids and related substances.[6] We have seen in Chapter 4 that peptides containing the two kinds (L and D) of amino acids could have been present on prebiotic Earth long before the appearance of authentic proteins. There is thus a possibility that such substances combined to form the first cellular envelopes, later to be joined by phospholipids and by true proteins, eventually to be replaced by the latter.

CELLULARIZATION: EARLY OR LATE?

In the opinion of many scientists, the formation of cells initiated the development of life. Some sort of physical aggregates or vesicles are believed to have formed first and to have served as seeds on which, or within which, chemical events of the kind sketched out in the two preceding chapters took place. In one version of this theory, these structures were associated from the start with light-sensitive molecules that allowed the biogenic process to be supported by sunlight energy right from its onset.[7]

Others, among them Miller and Eigen, already cited, have proposed, instead, that life arose in an unstructured "primeval soup" and that cellularization came later. Opponents of this theory point out that no conceivable "soup" could ever have been thick enough to harbor the complex chemical processes that led to life. To this criticism, Miller responds with his model of a "drying lagoon," in which evaporation would have produced the required concentration.

Gunter Wächtershäuser, a German chemist and patent lawyer who has developed a considerable interest in the origin of life, rejects both the primeval cell and the primeval soup in favor of a two-dimensional process unfolding on the surface of some submerged rocks before becoming confined.[8] He has proposed highly elaborate models of such a process, largely inspired by existing biochemical mechanisms.[9] Some of his ideas are being tested experimentally, with encouraging results.

It would be useful, in order to distinguish among these various theories, to have some information on the site where life originated. This point will be addressed later in this chapter. As will be seen, some suggestive clues are available, but they are not sufficiently unequivocal to be helpful. All that can be said for the time being, as shown in the preceding chapter, is that partitioning of nascent life into a large number of discrete entities became mandatory when molecular selection could no longer solely account for further development and had to give way to cellular selection. Whatever model is adopted for the origin of life, this stage must have been reached at the latest by the time the protein-synthesizing machinery began to settle into place and genes became selected on the strength, not of their molecular properties, but of the properties of their protein products. It is, of course, possible that cellularization occurred earlier. Much depends on the chemical nature of the first boundaries. If phospholipids or similar bilayer-forming molecules were required, this would argue in favor of a relatively late event, occurring at a stage where a certain degree of metabolic sophistication had already been achieved.

CELL DIVISION

The ability to divide must have been a property of protocells from the moment their existence became indispensable. The reason for this is simple. If, as pointed out earlier, protocells were required to allow a process of cellular selection, they could have played this role only if they could multiply.

This means that even very primitive protocells, just beginning to acquire RNA-dependent protein synthesis, must already have been able to divide. The mechanisms involved in this process were no doubt very crude, no more elaborate, perhaps, than the random kind of splitting that affects soap bubbles subject to physical stress, but sufficient, nevertheless, to allow a certain segregation of genes. Because of the importance of correct division for participation in Darwinian competition, it may be surmised that any genetic modification leading to a more reliable form of division would likely be retained by natural selection.

FROM PROTOCELLS TO THE FIRST CELLS

The first protocells were very different from even the simplest conceivable bacterial cell that could fulfill our description of the common ancestor of all life. As we have seen, they must, after their first appearance, have gone through a long developmental phase, during which genes and their protein products progressively lengthened up to their present size, leading to the production of more numerous and more efficient protein enzymes. The protocells must also at some time have switched from RNA to DNA, with all that such a shift implied in terms of new processes. Further progress would have included the adjoining of protein components to the ribosomal RNAs, with formation of the first ribosomes, construction of the first chromosome from the linking together of the genes, and better control of replication and transcription.

As the protocells thus evolved in the direction of increasing autonomy, there must have been a parallel evolution of their cellular envelopes, which must have acquired a variety of entry and exit ports and other complex systems adapted to the protocells' increasingly demanding metabolic requirements. At the same time, the mechanisms of cell division must have become increasingly organized and coordinated with DNA replication. Finally, organisms beginning to resemble present-day bacteria would have emerged.

THE WAY TO THE COMMON ANCESTOR

The end product of the long exploratory process just sketched out could have been the common ancestor, but there is no special reason why it should have been. If the estimates arrived at in the preceding chapter are anywhere near the truth, the actual genesis of viable bacterium-like organisms could have taken only a small fraction of the time that elapsed between the moment Earth became capable of harboring life and the age of the earliest fossil signs of life so far detected. It is thus possible that the common ancestor emerged only at the end of tens of millions of years of evolutionary history, punctuated by a wide diversity of intermediate forms of which none have been perpetuated in descendants that would allow us to trace the pathway followed.

We shall encounter this problem several times in our attempt to reconstruct the history of life. At some critical stage, there is a *bottleneck* that allowed only one successful form to evolve further, up to the present day. The impression is left either of a stringently channeled process or of some highly improbable, lucky breakthrough. In reality, this impression may be false. Before reaching the bottleneck, many gropings may have produced many forms vying for survival, until a particular trait gave one of the forms such a selective advantage under the prevailing conditions that all the others eventually disappeared. The common ancestor may have been such a lucky form, adapted to a set of particularly stringent environmental conditions.

THE CRADLE OF LIFE

From what we have seen in Chapter 3, there is no compelling reason for believing that life came to Earth from outer space. It thus seems reasonable to assume, until proven otherwise, that life started on our planet. But where on our planet? This question was briefly addressed in the beginning of Chapter 3, when the possible properties of the common ancestor were considered. Mention was made of the theory, entertained by a number of scientists, that life originated in a very hot environment.

The "hot-cradle" theory was first proposed in the 1980s by the American microbiologist Carl Woese, famous for the discovery that prokaryotes form two distinct evolutionary domains that separated at the dawn of microbial life (see Chapter 8). Woese's main tool was comparative sequencing (see Introduction and Chapter 7) of ribosomal RNAs. In addition to showing the early split of prokaryotes into two domains, his results indicated further that the most ancient microbes in both domains were thermophiles, that is, adapted to high temperatures.[10] These findings happened to coincide with the discovery of deep-sea hydrothermal vents, those abyssal volcanic springs already mentioned in Chapter 3. The possibility that life may have started in those mysterious, sulfurous depths has since attracted considerable interest.

There is much to be said for a volcanic cradle of life. Such an environment could have been a rich source of energy and of elements, such as sulfur, iron, and phosphorus, suspected of having played an important role in biogenesis.[11] Also, it seems likely that our young planet, which was just beginning to recover from the convulsions of its violent birth, was still erupting with volcanic outbursts all over its surface at the time life originated.

The hot-cradle theory has been criticized on the grounds that fragile chemical intermediates could not have survived such conditions long enough for the life-generating process to unfold productively. Some investigators, including Stanley Miller, the founder of prebiotic chemistry, and his former associate Jeffrey Bada, have proposed, on the strength of this argument, that life must have arisen at a very low temperature, near the freezing point of water. It could, however, be argued that heat accelerates all chemical reactions, not just the degradative ones. If fruitful reactions were favored by heat as much as the degradative ones, life could have started at any temperature. One could even visualize situations where heat might promote the development of life (if heat favors syntheses more than degradations). In any case, the existence of thermophilic organisms certainly proves that heat-resistant life is possible; the question is whether this property is an early or a late acquisition.

Other objections have been raised in recent years against the hot-cradle theory. One is technical. The significance of the sequencing results indicating that the most ancient bacteria were thermophiles has been questioned by some experts. Furthermore, it has been pointed out that, even if the common ancestor were to have been a thermophile, this would not necessarily mean that life originated in a hot environment. As we have seen above, it is very possible, if not likely, that the common ancestor arose from the first cells through a long evolutionary history. If such was the case, the nature of the environment occupied by the common ancestor is irrelevant to the conditions in which life first arose. Thermophilia could be a late property. It may even have been a saving one if heat created the bottleneck from which the ancestor emerged.

A question related to this problem concerns the source of the energy that supported the early biogenic processes. If, as some workers believe, light was involved in the primary process, only surface waters could have harbored the origin of life. There are some difficulties with this theory, however. The main one is that photosynthetic life is a discontinuous process, daily interrupted by hours of darkness. Present-day photosynthetic organisms live on their reserves during that time. It is difficult to visualize a primitive system endowed with such an ability from the start. Another difficulty is that photosynthetic systems, at least in their present form, depend on complex, membrane-embedded molecular associations.[12] Primitive systems of this sort are not readily imagined.

FINAL COMMENTS

This and the two preceding chapters were headed by the question: How did life arise? The answer to this question is clear: We don't know; and we may not know for a long time to come. But, at least, we can make conjectures that are open to experimental testing. This is because we understand life. This dependence of research and speculation on knowlege should be even greater in the future if the congruence argument is followed. The history of life, including its origin, is written into the very fabric of present-day organisms. It is up to us to decipher the fine print and draw the conclusions.

A major lesson we have learned, already emphasized several times in the preceding pages, is the primacy of *chemistry*. In many popular accounts, the respective roles of chemistry and information in the origin of life are presented as a "chicken-or-egg" problem. This is wrong. There can be no doubt that chemistry came first. Not just the kind of chemistry that produces small given molecules—mostly the province of cosmic chemistry—but the chemistry that involves complex interactions among these molecules and is supported by a constant flow of energy; in other words, *metabolism*. As shown in the preceding chapter, the long succession of events that must obligatorily be posited to account for the development of the first genes and the first protein enzymes could not possibly, whatever their mechanisms, have taken place without a solid protometabolic underpinning steadily maintained for as long as several millennia.

There is an important implication to this condition. By virtue of their chemical nature, the phenomena that gave rise to life were closely dependent on the physical-chemical conditions—lighting, humidity, temperature, acidity, availability of organic precursors, presence of mineral elements, etc.—that prevailed around them. What these conditions may have been is still, as we have seen, the subject of much debate. But one thing is clear. Whatever the conditions were, they must have enjoyed considerable *stability* in order to continue supporting the early chemical processes during the time required—at least several millennia, according to my estimate (Chapter 5)—for emerging life to be capable of withstanding significant environmental changes. It will be important for the geochemists who are trying to reconstruct the early history of our planet to identify environments that may have remained relatively unchanged physically and chemically for such a long duration. The stability condition, incidentally, is one more reason for believing that life arose "fairly quickly."

An exciting task awaiting future investigators will be to try to recreate the most probable prebiotic conditions in the laboratory, just as Miller did for the early atmosphere postulated by Urey, and to look for signs of incipient metabolism. There is also the hope that the conditions still exist in some pristine areas of our planet. Here, however, there is the risk of running into the celebrated "warm little pond" evoked by Darwin in a letter to a friend. Even if such a pond contained everything needed for life to appear, Darwin warned, the presence of existing life would stifle the process. "At the present day," he wrote, "such matter would be instantly devoured or absorbed." A natural cradle of life must perforce be sterile.

There is one more lesson to be learned from what we have seen. It concerns the role of *chance*, already examined at the end of Chapter 3. The argument was made that chance could have had little to do with the origin of life because chemical and, therefore, highly deterministic processes were involved and, also, because there must have been many steps. But what about after natural selection started playing a role, first, at the molecular and, later, at the cellular level? It is often argued—and I shall come back to this in subsequent chapters—that Darwinian selection is dominated by contingency because it operates on variations that are offered to it by purely accidental events. A key notion introduced in the preceding chapter is the likelihood that, at each critical step, chance provided natural selection with an essentially *exhaustive* array of possibilities to choose from, thus allowing for optimization in spite of contingency. This consideration, while invalidating the argument in favor of intelligent design, reinforces the conclusion that life was *bound to arise* under the physical-chemical conditions that prevailed at the site of its birth. It also supports the view that, if the same conditions should obtain elsewhere, life would likewise arise there, in a form closely similar in all its main features to life on Earth. These notions will be further developed in subsequent chapters.

7. The History of Life

O N 22 OCTOBER 1996, POPE JOHN-PAUL II, ADDRESSING THE
Pontifical Academy of Sciences, solemnly declared that "the
theory of evolution is more than a hypothesis." Coming from
an institution that took more than three centuries to rehabilitate Galileo
and, a bare 50 years ago, opposed publication of the attempts by the
French Jesuit Pierre Teilhard de Chardin to reconcile evolution with religion, such a statement underlines better than any scientific argument
the compelling character of the proofs of biological evolution. This is
now a solidly established fact, which is contested only by unyielding
ideologues. What remain matters for discussion are the details of the tree
of life, as well as the mechanism of evolution.

BIOLOGICAL EVOLUTION IS A FACT

The notion that life may have a history dates back only little more than
two centuries. Before that, living species were viewed as given once and
for all. Life had no more history than the universe. Only we, humans,
had a history. All the rest, Sun and stars, continents and oceans, plants

and animals, formed the immutable infrastructure created to serve as setting and support for the human adventure. That this idea might be wrong was first suggested by the fossils.

FOSSILS HAVE REVEALED THAT LIFE HAS A HISTORY

It took a long time for the significance of fossils to be correctly recognized. They were first seen as the remnants of living beings that no longer existed in our countries but still subsisted in unexplored parts of Earth. Interpretations even went so far as to accept that species may have disappeared completely, victims of the Flood, for example. But the possibility that new species could have arisen from more ancient species made its way only very slowly. Thus, the celebrated French naturalist Georges Cuvier, who flattered himself to be capable of reconstituting the entire skeleton of an extinct animal from a single fossil bone, remained until his death in 1832 a fierce opponent of evolutionary theories.

It needed the development of geology for the notion of evolution to become progressively evident. It slowly came to be accepted, though not without a great deal of hesitation and debate, that Earth has a history whose archives are recorded in the geological strata. Observers were then struck by the fact that the more ancient terrains contained only fossils of primitive living beings, to which were added increasingly advanced forms as the terrains became younger. This led to the so-called transformist hypothesis, according to which more complex organisms arose from the transformation of simpler organisms. This hypothesis, already prefigured in the writings of the French philosopher of the Enlightenment Denis Diderot, was proposed at the turn of the eighteenth century more or less simultaneously in France by Jean-Baptiste de Monet, chevalier de Lamarck, and in England by Erasmus Darwin, the grandfather of the famous Charles. Contrary to popular belief, it is not Charles Darwin who discovered biological evolution. His merit lies in his having proposed *natural selection* as the mechanism for this phenomenon. I shall come back to this point.

The theory of evolution was strongly disputed at first, notably by many religious bodies, who took it as an attack on their beliefs. In attempts to refute it, the argument was made that God could have created living beings in successive stages, without there necessarily being a relationship between them; or else that He simply created the world as it is, in one single shot. This is what is still claimed today by strict creationists, who assert that God created the world a little over 5,000 years ago, with all its present-day components, including living beings, fossils, geological strata, and even partially disintegrated radioactive elements bearing false witness to a very ancient history. The image of a God who would thus find pleasure in misleading humankind—to test its faith?—does, however, strain credibility. As Einstein once said, "the Lord is subtle, but not malicious." It has already been mentioned that the Catholic Church, long opposed to the notion of evolution, has recently bowed before the evidence of facts.

The existence of biological evolution began to be accepted in scientific circles in the course of the second half of the nineteenth century under the influence, notably, of the monumental work of Charles Darwin. One of its staunchest defenders was the German biologist and philosopher Ernst Haeckel, remembered for his so-called recapitulation law, according to which "ontogeny recapitulates phylogeny," by which he meant that advanced living beings, in the course of embryological development (*ontos* comes from a Greek word meaning what is), go through the main stages of their evolutionary development (*phylon* means race in Greek). Thus, a human embryo starts as a single cell, then gives rise to formations that recall the first invertebrates, subsequently goes through a fish state, with gill-like structures, to finally become mammal and human. There is a kernel of truth in this notion, which, however, no scientist nowadays is ready to grace with the status of "law."

Haeckel was the first to construct what is known as a *phylogenetic tree*, joining species according to kinship, as in our genealogical trees. To accomplish this, Haeckel had to add a solid dose of imagination to the meager data he had available. Even today, the enterprise would be highly hazardous if only fossils were available, be they illuminated by the recapitulation notion.

Indeed, fossils, as historical documents, provide only fragmentary information. They are mostly bones, teeth, shells, and other mineral or mineralized vestiges, or, alternatively, imprints of plants, feathers, footsteps, crawling, and other tracks. Nothing remains of the cellular structures, the enzymes, the genes (except for recent fossils, which have yielded traces of DNA sufficient for amplification by modern techniques), nor of the other molecular characteristics of the organisms that have left these vestiges. Especially, nothing at all remains of the organisms of which no fossil remnant is known, which means the majority of living beings that have followed each other on the planet. Indeed, barring a few bacterial traces (see Chapter 3), very few fossils older than 600 million years are known, whereas life is more than six times more ancient. Enormous gaps would therefore be left in the history of life as it can be reconstructed had not molecular biology offered a new, extraordinarily productive approach.

THE HISTORY OF LIFE IS WRITTEN
IN PRESENT-DAY LIVING BEINGS

The principle of the new technology has already been illustrated in the Introduction. It is based on the comparison of the sequences of information-bearing molecules—DNAs, RNAs, or proteins—from different organisms in which the molecules, called homologous for this reason, carry out the same function. First applied to proteins and later to RNAs, this technique now rests almost exclusively on the sequencing of DNA molecules[1], thanks to the development of an extraordinarily powerful collection of tools for the isolation[2] and sequencing[3] of this substance. Once the sequence of a DNA is known, those of its transcription (RNA) and translation (protein) products are readily reconstituted RNA simply by replacing T with U in the complementary strand, and protein with the help of the genetic code.[4]

In order to examine more closely the possibilities and limitations of comparative sequencing for phylogenetic tree building, let us look once again at the five sequences shown in the Introduction:

Colibacillus:	ILDIGDASAQELAEILKNAKTILWNGP
Wheat:	GLDIGPDSVKTFNDALDTTQTIIWNGP
Fruitfly:	GLDVGPKTRELFAAPIARAKLIVWNGP
Horse:	GLDCGTESSKKYAEAVARAKQIVWNGP
Human:	GLDCGPESSKKYAEAVTRAKQIVWNGP

It will be remembered that these sequences represent, each letter standing for an amino acid, the same 27-amino acid stretch in a protein enzyme (phosphoglycerate kinase, or PGK) extracted from the different organisms mentioned. My Brussels colleagues Fred Opperdoes and Pol Michels, who have kindly provided me with the above sequences, have similarly investigated 14 other organisms, including mice, yeast, fungi, and a number of protists and bacteria. But the five sequences shown suffice to illustrate the main points.

First, there are the eight positions (in bold) occupied by the same amino acid in all five sequences. Five of these positions are actually constant in all nineteen organisms investigated, and two are different in only one. As demonstrated in the Introduction with the help of the typing-monkey allegory, these similarities can be explained only on the assumption that the invariant positions are conserved, that is, were derived unchanged from the same ancestral sequence, thus providing incontrovertible proof of the single evolutionary origin of all the organisms concerned.[5] As to the differences, they were attributed to changes, or *mutations*, occurring in the course of evolution, and they were used to connect the human, horse, and fruitfly sequences in a simple tree based on the hypothesis that the number of differences between two evolutionary lines is a measure of the time that has elapsed since the lines diverged from their most recent common ancestor. Thus, with the human sequence showing two differences with respect to the horse sequence and twelve with respect to the fruitfly sequence, the conclusion was drawn that the human/fruitfly bifurcation antedates the human/horse bifurcation by a time factor of six. Such, explained in a highly schematic fashion, is the principle of *phylogenetic tree building by comparative sequencing*.

Note that the result would have been much less satisfactory if I had tried to include the other two species in the tree, with colibacillus (13 differences) and wheat (14 differences) hardly more distant from humans than fruitflies (12 differences), which doesn't make sense. This is not because the technology is at fault, but simply because the sampling is too small. In actual fact, when the complete sequences, which contain more than 400 amino acids, are compared for all 19 investigated species, as was done by my colleagues, the resulting tree turns out to have a perfectly acceptable shape.

Let us now consider mechanisms. What the sequence differences reveal are the consequences of *point mutations*, that is, genetic changes that have led to the replacement of one amino acid by another.[6] These mutations have obviously not caused the affected organisms to be eliminated by natural selection. Many of them may not, either, have had much to do with the survival of the organisms, since the molecules concerned presumably were perfectly functional before the changes occurred. As far as can be known, the mutations are *neutral*, or near-neutral. They are accidental changes that were retained, not by natural selection, but by *genetic drift.*[7]

In contrast with the observed replacements, the conserved positions most likely remained unchanged throughout evolution because they could not undergo any modification without the enzyme being impaired to an extent that caused the owners of the mutated gene to be eliminated by natural selection. Intermediate situations are possible. It could be, for example, that a previously unmodifiable position becomes open to change after the molecule has suffered some other modification that neutralizes the deleterious effect of this change.

Phylogenetic tree building by comparative sequencing is not nearly as simple as might be gathered from the proposed example. Mutations undergone by genes in the course of evolution are not just point mutations. They also include duplications, suppressions, inversions, and other rearrangements, often involving whole blocks of nucleotides. Successive modifications may correct each other, suggesting a closer kinship than there is in reality. It may even happen that DNA fragments containing one or more genes migrate horizontally (see Chapter 3) from a given organism to a very distant relative, thus inextricably muddling the trails. There is increasing evidence (see Chapter 8) that such transfers occurred on a large scale in the early stages of bacterial evolution. In addition, there is every reason for believing that the mutation rate is not the same for all parts of a genome, nor constant in time. Finally, it is obvious that only mutations that have not been eliminated by natural selection are available for this kind of exercise.

For all these reasons, the construction of phylogenetic trees by comparative sequencing is subject to numerous difficulties and uncertainties, which experts try to minimize by increasingly sophisticated algorithms and computer programs. These defects are, fortunately, compensated for by the fact that organisms possess thousands of different genes, of which each can, theoretically, be the object of an independent analysis. Multiple cross-checks are thus possible, allowing increasingly trustworthy reconstructions of the tree of life.

A major advantage of the new technology is that it can be applied to any extant organism. Phylogenetic trees now embrace the whole history of life, not only those few episodes that have left some fossil trace, not always easy to interpret at that. The main weakness of the technology is that it is indirect and rests on a number of assumptions. On the whole, whenever comparisons were possible, molecular phylogenies have been found to agree fairly well with those established with the help of fossils. This agreement is reassuring with respect to the validity of both methods and, in addition, has allowed molecular data to be put on an absolute time scale provided by paleontological data. Some discrepancies have been observed and traced to defects of either one or the other method. Especially, a rich crop of new data has been gathered and continues to be gathered almost daily. As of now, we already have available on the history of life an impressive amount of new information, of which some turned out to be totally unexpected.

Independently of those details, of interest only to experts, the deciphering of molecular sequences has unassailably established the fact of biological evolution, the human species included. It is astonishing, in the face of such glaring evidence, that large groups still exist that deny this fact on the strength of texts written more than 2,000 years ago. Even more astonishing, among those groups are many highly educated persons, including writers, journalists, lawyers, politicians, business people, engineers, teachers, philosophers, theologians, and even a few scientifically qualified individuals.

THE DRIVING FORCE OF BIOLOGICAL EVOLUTION

The notion of evolution is intimately linked to that of heredity. To be sure, the first evolutionists had only the vaguest ideas on this subject. They not only had no knowledge of DNA, but even the concept of a gene was foreign to them. It was only in 1866 that the founder of genetics, the Austrian monk Gregor Mendel, published the results of his patient observations on the crossbreeding of peas. And it took more than another 30 years before the scientific world rediscovered the now famous "Mendel's laws" in the obscure journal in which his works appeared. Long before that, however, biologists and, even, the common run of people obviously knew that children resemble their parents and, in a more general manner, that species perpetuate themselves. Humans give birth to humans, mice to mice, snails to snails, oak trees to oak trees, and so on. For evolution to take place, some element of variation had to be inserted into this continuity, so that eventually a primitive frog might emerge from a fish, or a human being from a monkey. Two notions have vied to explain the variation of hereditary characters.

LAMARCK AND THE HEREDITY OF ACQUIRED CHARACTERS

In Lamarck's view, the source of variation was usage. He proposed, as a mechanism of evolution, the heredity of acquired characters. Thus, he believed that giraffes owed their long neck to the stretching efforts of generations of giraffes trying to reach the higher branches of trees, each millimeter gained in this fashion becoming hereditary. For Charles Darwin, the long neck of giraffes is the result of accidental hereditary changes that turned out to have consequences useful for the survival and multiplication of the modified individuals.

We know today that Darwin was right. But his vindication did not come easily. In the beginning of this century, an Austrian biologist, Paul Kammerer, an ardent adept of Lamarckism, made headlines with experiments on a species of toads that copulate on land and whose males do not possess on their legs the "nuptial pads" that allow those that copulate in water to grip the slippery body of the female. Kammerer claimed that those pads appeared in terrestrially copulating male toads artificially forced, generation after generation, to copulate in water. Accused of having falsified his experimental results, Kammerer committed suicide. As late as 1971, Darwin notwithstanding, the British writer of Hungarian origin Arthur Koestler devoted an entire book (*The Case of the Midwife Toad*) to the defense of Kammerer's memory and of the authenticity of his observations.

On a much more dramatic level from a historical point of view, the apparent agreement between Lamarckism and Marxism allowed the Soviet agronomist Trofim Lysenko to impose himself for more than 30 years upon the Stalinist leaders, with consequences that, besides the persecutions of which numerous Soviet biologists were victims, were disastrous both for the development of genetics and for the success of agriculture in his country.

Lamarckism is abandoned today. Its death knell was sounded by the advances of molecular biology, in particular by the unidirectionality rule known under the name of "central dogma." We have seen (Chapter 2) that a modified protein cannot transfer this information to the gene coding for its sequence. There can therefore be no heredity of acquired characters. This affirmation needs, however, to be slightly qualified, in the sense that it applies exclusively to the classical type of heredity, that depending on DNA (or possibly RNA). It does not concern the cases, admittedly rare but nevertheless real, of the direct transmission of shape—of a prion, for example, or of a membrane—of authentically Lamarckian nature (see Chapter 2).

DARWIN AND NATURAL SELECTION

It was upon his return from his long, historic voyage on board the *Beagle* (1831–1836), with its much-publicized stop in the Galapagos Islands, that Charles Darwin started to elaborate the theory that was to make his fame. It took him more than 20 years to mature his ideas, always deferring their publication to a later date. He finally resolved to publish them in 1859, in his celebrated book, *On the Origin of Species by Means of Natural Selection*, after receiving a letter in which, in 1858, his friend Alfred Russel Wallace sketched out a similar theory. While keeping an honorable place for Wallace, history has retained—justifiably—the name of Darwin as author of the theory of natural selection.

Three elements, in addition to his many personal observations, inspired Darwin. First, there was the notion of evolution, already defended by his grandfather, Erasmus Darwin, who, however, believed in the heredity of acquired characters, like the French author of the transformism theory, Lamarck.

Then there was the concept of struggle for life, which Darwin drew from his reading of the famous *Essay on the Principle of Population as it Affects the Future Improvement of Society* by Thomas Robert Malthus, published for the first time in 1798 but still much appreciated in Darwin's time. In this book, the British economist, who has given his name to Malthusianism, put forward the theory that the geometric growth of populations must necessarily bring about an increasingly fierce competition for arithmetically growing resources. Hence the subtitle Darwin gives to his book: "*The Preservation of Favored Races in the Struggle for Life.*"

Finally, Darwin was struck by the numerous instances of artificial selection, which show that it is possible, by appropriate breeding, to exploit the natural variability of populations in order to favor one or the other hereditary trait. In this way, one can, he notes, obtain plants that are cultivated for their leaves, like cabbage, or for their flowers, like cauliflower, horses adapted to running or to pulling, dogs that track down game or retrieve it, sheep whose wool lends itself better to weaving rugs or cloth, and so on. The whole history of agriculture and animal breeding rests on this kind of exploitation.

It sufficed to combine those three notions to imagine that species owe their origin to the natural selection of the fittest among the diverse varieties engaged in the struggle for life. "How could I have been so stupid as not to have thought of it!" exclaimed the naturalist and likewise great voyager Thomas Henry Huxley, when he first read the *Origin of Species*. It was the same Huxley who, in 1860, in the course of a memorable confrontation with the bishop of Oxford, Samuel Wilberforce, who had asked Huxley whether he claimed to descend from monkeys by his grandfather or his grandmother, answered: "I would rather have a miserable ape for a grandfather than a man highly endowed by nature and possessed of great means and influence, and yet who employs these faculties and influence for the mere purpose of introducing ridicule into a grave scientific discussion." In 1932, at the dawn of Nazism, his grandson Aldous Huxley was to denounce, in his *Brave New World*, the perverse effects of artificial selection practiced on humans.

All that has been learned since the publication of the *Origin* has confirmed and reinforced the correctness of Darwin's theory. The advances of genetics and molecular biology have given a solid body to the vague notions of heredity and variability he and his contemporaries had to be content with. Today, we speak in terms of DNA replication and mutations, and we understand the mechanisms involved. The result is what is sometimes referred to as the *synthetic*, or *neo-Darwinian*, theory of evolution.

MUTATIONS PROPOSED, NATURAL SELECTION DISPOSES

The most important conclusion to emerge from our newly gained understanding is that hereditary variability, which is the foundation of all selection, whether natural or artificial, is entirely independent, as Darwin already suspected, of any sort of influence that might cause a defined hereditary change aimed at a desirable effect. All that we know of the molecular nature of the hereditary modifications, or mutations, responsible for this variability supports the view that the phenomena involved are *accidental products of chance* devoid of any intentionality. Which obviously does not mean that they are devoid of causality.

Many causes of mutations are known. First, there are all the mistakes and imperfections that can affect DNA replication and other natural forms of DNA processing, for example, the exchange of pieces of chromosomes, or crossing-over, that takes place in the course of germ cell maturation in plants and animals. Then there are all the physical and chemical agents that damage DNA. The former include ultraviolet light and all other forms of radiation of higher energy, such as X-rays, radioactive emanations, and cosmic rays. Among the latter are free radicals, such as arise from oxygen, and, especially, the numerous so-called *mutagenic* substances, of which most have been created by our industrial and commercial chemistry. Biological mutagenic agents also exist—for example, certain viruses that disorganize the DNA of the cells they invade.

These phenomena are now being studied intensively, under the pressure, notably, of environmental defense groups. Not only does the future cause worry, to the extent that mutations imperil the survival of species. There is immediate concern also for human health, because mutagenic agents are *carcinogenic*. From Hiroshima to Chernobyl, tests have been aimed at bringing to light an abnormal frequency of leukemias, thyroid cancers, and other pathologies, in the populations exposed to radiation. Even the surroundings of nuclear power plants, where no apparatus can detect any excessive level of radiation, are not spared by the "green police." Similarly screened are hosts of substances used as fertilizers, pesticides, food additives, etc. A whole risk evaluation industry has been spawned by such undertakings.

Mutagenic agents are sometimes deliberately exploited in attempts to create useful variants for some application or research. The first makers of penicillin used this strategy. When, in the course of the Second World War, the antibiotic was first extracted industrially from cultures of *Penicillium notatum*, yields with natural strains of the mold did not exceed five units per cubic centimeter of the culture medium. The cost was exorbitant. In order to improve matters, mold samples were exposed to strong doses of X-rays—a strategy reminiscent of "carpet bombing"— and the surviving mutants were screened for their penicillin-producing ability. It proved possible, by this simple means, to increase the penicillin yield of the cultures more than 20-fold and thus to decrease the cost of the miracle drug by the same factor.

Today, this procedure has become a choice tool for the analysis of many natural phenomena. Instead of a frontal attack on the phenomenon, which is often too complex to lend itself to a simple experimental approach, the cells or organisms under study are "bombarded" with radiations or mutagenic substances. A search among the victims then fishes out individuals showing an anomaly of the phenomenon one is interested in. With the help of molecular biology techniques, the affected gene is then identified, isolated, amplified, sequenced, translated, either in reality or on paper (with the help of the genetic code), and finally, if all goes well, characterized functionally. Numerous advances have been made in this manner, notably in our understanding of embryological development and of immune defense against cancer.

Whether natural or artificial, mutations all have in common that they are *unrelated to any sort of intentionality*. No natural mechanism is known whereby a mutation, however useful, can be elicited otherwise than by accident. Even an acquired advantage, such as a protein modified by usage, cannot, as we have seen, be perpetuated in DNA by a corresponding mutation. The "carpet bombing" technique is based on recognition of this fact. Unable to aim, we attack at random with a large number of projectiles, in the hope of hitting the target. Actually, this is no longer entirely true. It is now possible, by appropriate manipulations, to substitute one base for another in a defined site of a DNA sequence (*site-directed mutagenesis*). But site-directed mutagenesis is a human invention. In nature, all mutations, including those that have shaped the course of evolution, are accidental events. This is a keystone of modern neo-Darwinian theory.

It is from this natural variability, constantly maintained by the accidents of mutations, that, for the last 10,000 years, selectors have derived, first blindly, then by increasingly sophisticated empirical methods, all the plants we grow in our fields, meadows, parks, and gardens, all the animals we breed to feed us, carry us, work for us, and keep us company. It is from the same variability that, for almost four billion years, natural selection has derived the species that have succeeded each other on Earth and inhabit it today.

There lies Darwin's brilliant intuition. From the moment there is competition for existing resources, species genetically disposed to produce the most vigorous and abundant progeny must perforce outnumber those that reproduce less readily. This is so evident that some accuse Darwinian

8. The Invisible World of Bacteria

FOR NEARLY THREE BILLION YEARS, LIFE WOULD HAVE BEEN VISIBLE only through its effects on the environment and, sometimes, through the presence of colonies, such as stromatolites (see Chapter 3), associating trillions of microscopic individuals in formations that could have passed for rocks were it not for their sticky surface and changing colors. The whole panoply of plants, fungi, and animals that now covers the terrestrial globe with its splendor did not exist. There were only unicellullar organisms, starting almost certainly with bacteria.

THE BACTERIAL WAY OF LIFE

To most of us, the word "bacteria" conjures up specters of plague, cholera, diphtheria, tuberculosis, and all the other scourges from which we were delivered by the genius of Pasteur and his contemporaries. But pathogenic bacteria are only a small minority. Most bacteria are harmless. Many are useful, even indispensable, for the economy of the living world. There are bacteria everywhere, adapted to an enormous variety of different environments. Among their many beneficial actions, bacteria are, together with a few mushrooms and molds, largely responsible for the decomposition of dead plants and animals, which ensures the recycling of carbon, nitrogen, and other biogenic elements.

The great diversity of bacteria is constructed according to a common plan, which seems to be the simplest compatible with independent life. All consist of single cells. These may form colonies, sometimes complex, but no veritable pluricellular organisms. Bacterial cells are generally of small size, on the order of one-thousandth of a millimeter. They are surrounded by a membrane, most often protected by an external rigid wall. The mineralized remnants of that shell make up the microfossils mentioned in Chapter 3. Frequently, no special structure can be discerned inside bacterial cells. In particular, their genome is not contained within an enclosure, as it is in the nucleus of plant and animal cells. Hence the name of *prokaryotes* by which they are known (see Chapter 1).

PHYLOGENY OF BACTERIA

For a long time, prokaryotes were believed to form a single group from which the much more advanced *eukaryotes* were assumed to have arisen at a relatively late stage of evolution. Then, in the late 1970s, the American microbiologist Carl Woese[1] came up with startling findings that shattered this view.

THE THREE DOMAINS

Using comparative sequencing of ribosomal RNAs (see Chapter 7), Woese found that the prokaryotes are divided by a deep gap going back to the earliest origins of unicellular life. In one group, which he called *eubacteria*, were all the familiar pathogenic and nonpathogenic microorganisms traditionally studied by bacteriologists.[2] The other group, to which Woese gave the name *archaebacteria*, because he believed them to be particularly ancient, comprised the vast family of methane-producing organisms, or methanogens, which are found in a wide variety of oxygen-poor milieus, as well as a number of highly specialized kinds, now grouped under the name "extremophiles," adapted to very hot (thermophiles), very acid (acidophiles), or very salty (halophiles) environments. Many other forms adapted to milder habitats have since been added to the archaebacterial family.

Another surprising conclusion arising from Woese's analyses was that the separation between prokaryotes and eukaryotes seemingly took place much earlier than generally assumed, almost at the same time as the divergence of the two prokaryotic families. It thus appeared that the three groups originated more or less simultaneously from the common ancestor, forming what Woese considered true distinct kingdoms, for which he has since proposed the term "domain." To stress this distinction, he has renamed eubacteria *Bacteria*, archaebacteria *Archaea*, and eukaryotes *Eucarya*. This terminology has gained acceptance, in spite of the possible confusion between Bacteria and bacteria, a term still widely used in nontechnical publications—including this book—as a synonym of the word "prokaryotes," virtually unknown by the general public.

THE ROOTING PROBLEM

A major problem created by these remarkable findings has been the rooting of the tree. An evolutionary "trifurcation," leading simultaneously to A (Archaea), B (Bacteria), and E (Eucarya), can hardly be contemplated and must hide two successive bifurcations. There was thus the possibility of the first bifurcation leading from the common ancestor to A and B, with E branching later from one or the other of those two lines. Or there was the alternative eventuality, seemingly more unlikely but nevertheless entertained by a number of investigators, of E arising from the first fork together with either A or B, and of the second prokaryotic line emerging later from one of the two arms of the first fork. At first sight, one would have expected this problem to be rapidly solved as more genes were sequenced in the three domains. On the contrary, the more data, the murkier the situation.

One reason for this is that very ancient events, such as those one attempts to reconstruct, lend themselves poorly to molecular phylogenies because the original trails are largely erased by the many genetic changes that have piled up in the course of time. Another, particularly important, cause of confusion is *horizontal*, or lateral, gene transfer. In opposition to the usual, vertical kind of gene transfer, which occurs from generation to generation in the course of reproduction, this phenomenon, already mentioned earlier (see Chapter 3), consists in the transfer of DNA from one bacterial cell to another. Several mechanisms whereby this can happen are known. The fragments of DNA transferred in this way may be of considerable size, and the donor and acceptor cells may be totally unrelated. To give just one example of this phenomenon, it has been found that some thermophilic eubacteria, which happen to be among the most ancient representatives of this group, have gained up to 16 to 24 percent of their genes from thermophilic archaebacteria.[3] This fact suggests that some early eubacteria may have acquired the ability to withstand a hot environment by the horizontal transfer of the relevant genes from heat-resistant archaebacteria.

Such cases, of which many instances are known, obviously complicate enormously the construction of phylogenetic trees from comparative sequencing data. Many investigators have practically given up the image of a typical tree emerging from a single root, replacing it with that of a network of interconnected ramifications springing from the common ancestor in a manner that may be impossible to reconstruct. We shall see later (Chapter 10) that the problem is particularly intractable as regards the origin of eukaryotes.

The invasion of primitive Earth by the invisible world of bacteria has had consequences of crucial importance, of which two, in particular, deserve mentioning. First, bacteria have, by their chemical activities, deeply modified the conditions prevailing on our planet, to the extent of rendering it totally unrecognizable to a hypothetical observer who had known it before the appearance of life. The most revolutionary among these effects has been the production of molecular oxygen, which almost certainly was absent from the early atmosphere and reached its present level of 21 percent of the terrestrial atmosphere relatively late, essentially as a result of biological activity. The second historic innovation of the bacterial era, already alluded to above, has been the remarkable evolutionary phenomenon that has given birth to the first eukaryotic cells, ancestral to the whole visible living world. The next two chapters will be devoted to this epoch-making event.

Photosynthetic Bacteria Presented Life with a Poisoned Gift, Now Vivifying: Oxygen

We saw in Chapter 1 how aerobic (living in air) organisms derive energy from the oxidation of foodstuffs. In such reactions, atmospheric oxygen (O_2) is used to convert the carbon and hydrogen atoms of organic substances into carbon dioxide (CO_2) and water (H_2O), respectively.[4] The most important such reactions are coupled to the energy-requiring assembly of ATP from ADP and phosphate. The reverse of this assembly, the energy-yielding splitting of ATP to ADP and phosphate, powers all the main forms of biological work, thereby harnessing, by way of ATP, the performance of work to the energy released by oxidations. Without oxygen to support the oxidative phosphorylation of ADP to ATP, we quickly suffocate, as do all other aerobic organisms under the same conditions, unable to carry out vital functions, such as breathing, heartbeat, and conscious thought, that require energy derived from the splitting of ATP.

AN OXYGEN-FREE PRIMEVAL WORLD

Surprisingly, the crucially important exploitation of oxygen for the production of energy is a late development in the history of life. All the available geochemical evidence indicates that the terrestrial atmosphere was initially almost totally *devoid of oxygen* and remained so until about 2.3 billion years ago. The level of atmospheric oxygen started rising from that time on, reaching a value compatible with aerobic life a few hundred million years later. Until this happened, all living organisms were *anaerobic* (living without air) and derived their energy from chemical reactions, coupled to the assembly of ATP, that did not use oxygen.[5]

THE BIOLOGICAL PRODUCTION OF OXYGEN

The phenomenon responsible for the appearance of oxygen in the atmosphere is biological *photosynthesis*, more precisely the advanced form of this process, such as is carried out by green plants in today's world. As was mentioned in the first chapter, this process reverses, with the help of light energy, the kind of energy-yielding reactions that take place with oxygen. In particular, carbon dioxide and water are converted into a sugar of schematic formula $C_6H_{12}O_6$, or $(CH_2O)_6$. A bit of arithmetic reveals that this reaction produces one molecule of free oxygen for every molecule of carbon dioxide and water used.[6]

There is every reason to believe that oxygen first appeared and accumulated in the atmosphere thanks to this process and that the organisms responsible for this major change in our planet's chemistry were relatives of extant photosynthetic bacteria formerly known as blue-green algae, and now called *cyanobacteria* (from the Greek *kyanos*, blue). These organisms, which, as we shall see (Chapter 10), gave rise to the chloroplasts, the photosynthetic organelles of green plants, could be of very ancient origin.

When exactly in the history of life on Earth cyanobacteria first appeared is a matter of much current debate. As mentioned in Chapter 3, the view that cyanobacteria were already present as early as 3.5 billion years ago has been strongly defended by Schopf on the basis of traces he believes to be fossil remains of such bacteria. This contention is not readily reconciled with the geochemical data indicating that the atmospheric oxygen content only started rising more than one billion years later. It has been suggested, to account for this major discrepancy, that the oxygen produced by early photosynthesis was trapped by ferrous iron (Fe^{++})[7], which was abundantly present in the oceans, generating insoluble deposits rich in ferric iron (Fe^{+++}). Allegedly attesting to this phenomenon are massive, iron-rich, geological structures, called "banded iron formations" because of their striped appearance, whose age indeed covers the period from 3.5 to 2.0 billion years ago. Only when this iron "oxygen sink" became saturated, so the explanation goes, did the oxygen generated by photosynthesis begin to accumulate in the atmosphere. This explanation is plausible but, perhaps, no longer necessary. We have seen in Chapter 3 (note 3) that the cyanobacterial origin of the traces discovered by Schopf is now seriously being questioned. Whatever the outcome of this controversy, the critical time is around 2.3 billion years ago. It is then that, according to all available data, the level of atmospheric oxygen started rising. It is believed that this event caused catastrophic perturbations in the biological equilibria of the time.

THE OXYGEN CRISIS

Oxygen is such a vital element for us that we find it hard to imagine that it once raised the greatest menace life ever faced. Yet, strictly anaerobic microbes are known, for which oxygen is a deadly poison. The case of the bacillus of gaseous gangrene was mentioned in Chapter 1. Oxygen is not toxic by itself but it generates toxic derivatives when in contact with certain biological substances. These derivatives include the superoxide ion (O_2^-), the hydroxyl radical (OH), and hydrogen peroxide (H_2O_2). All aerobic organisms possess systems that render these derivatives harmless.[8] Some even take advantage of the derivatives in offensive actions. Our white blood cells, for example, attack pathogenic bacteria with oxygen-derived free radicals. Most organisms do much more than simply defend themselves against oxygen. As we have seen, they utilize this gas to carry out the metabolic combustions on which they depend to satisfy their energy needs. For those organisms—including ourselves—oxygen, far from being a poison, is a substance of vital importance.

Born in the absence of oxygen, the first living organisms, and all those that succeeded them until about two billion years ago, most likely resembled present-day strict anaerobes in their sensitivity to oxygen. It is suspected that many succumbed when oxygen started rising in the atmosphere. The only ones to survive what is sometimes called the "oxygen holocaust" were those that found shelter in some oxygen-free niche and those that acquired in time the necessary means of defense. To begin with, these means were simply enzymes capable of neutralizing the poisons created by oxygen.[9] A crucial step was accomplished with the generation of the first system by which oxygen consumption is *coupled* to the assembly of ATP. This required a special adaptation of the cell membrane of unknown origin. Once this step was accomplished, increasingly complex and efficient ATP-generating oxidizing systems were assembled in the membrane, and the organisms became increasingly dependent on oxygen for their energy supply. The crowning event in this process was the appearance of highly perfected aerobic bacteria, from which, as will be seen in Chapter 10, the main oxidizing systems of eukaryotes are derived.

BACTERIA, SUPERSTARS OF THE LIVING WORLD?

Related to the first forms of life, bacteria have been around for close to four billion years, unaccompanied during a good part of that time by any other form of life. Should conditions deteriorate, bacteria would still be present long after all so-called higher forms of life, ourselves included, have disappeared. Infinitely adaptable, almost indestructible, they are everywhere, thriving in even the most inhospitable environments: the ice of polar caps, the boiling water of volcanic springs, the drying brine of salt lakes, sulfuric acid, caustic potash, discharges loaded with lead, mercury, or almost any other natural or artificial pollutant, even naked rocks buried several kilometers down in the depths of Earth's crust.

All these qualities, according to some contemporary biologists, place bacteria high above eukaryotes, which these unconditional admirers of prokaryotes tell us are nothing but associations of bacteria. We shall see later what is to be thought of this view, which is notably defended by

9. The Mysterious Birth of Eukaryotes

THE PROBLEM

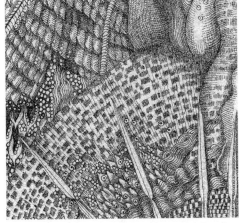

WHAT IS COMMON among a yeast, a diatom, a tobacco leaf cell, and a human neuron? At first sight, even with the help of a good microscope, not much. To be sure, all four are cells, and they have a nucleus, which puts them in the category of eukaryotes, as opposed to prokaryotes, or bacteria. But, otherwise, the differences among them are enormous. The yeast cell is small, the size of a big bacterium, and it shows few inner components. The diatom is big and complex; especially, it is surrounded by an elaborate mineral shell of exquisite design.[1] The leaf cell, which inhabits a small, rigid cellulose chamber (Hooke's original cell, see Chapter 1), is conspicuously filled with green particles. As to the neuron, its most striking feature is represented by remarkable extensions, some exceedingly slender and amazingly long—up to one meter or more—others branching into a dense bush of ramifications.

Yet, modern electron microscopy has revealed within all four cells a complex set of inner structures that are the same in all four and are not observed in bacteria. Thanks to advances in biochemistry, the functions of these structures, also the same in all four cells, have been unravelled. Finally, with the progress of molecular biology, many of the genes behind the common structures and functions have been sequenced and found to be related in all four cell types. In fact, all available evidence indicates that the four cells considered are, together with all other eukaryotes, the descendants of a *single ancestral form*. This form should not be confounded with the common ancestor of all life referred to in the beginning of this book (see Introduction and Chapter 3). As will be pointed out

later, we are dealing here with a new and much later evolutionary bottleneck, from which are issued all *eukaryotic* organisms, to wit: a collection
of unicellulars grouped under the name of protists and the three great
groups of pluricellulars, the plants, fungi, and animals, including human
beings.

Eukaryotic cells share all the basic properties of prokaryotes. They are
made of the same kinds of chemical components, use similar metabolic
pathways, and depend on the same genetic mechanisms. But they are
much more voluminous than prokaryotes, and their internal organization
is so much more complex that their origin from bacterial ancestral forms
seems almost incredible. This, nevertheless, is what must have happened,
as there is abundant proof, irrefutably imprinted in their molecular characteristics, that the two are related.

No fossil trace of this momentous transformation has yet been uncovered. Nor, as far as is known, has any of the intermediates the transformation must have involved left a legacy that has stretched unto this
day, thus allowing its characterization. We have only our knowledge of
the two cell types to attempt a reconstruction of the pathway between
them. Before embarking on such an attempt, which will be the subject
of the next chapter, let us take a look at what is known.

THE DISTINCTIVE FEATURES OF EUKARYOTIC CELLS

Eukaryotic cells, even the simplest of them, are *organisms* of considerable
complexity, composed of many different parts endowed with distinct
functions. In the description that follows, I have restricted myself to the
properties that are common to the vast majority of eukaryotic cells, keeping details to the minimum required for a meaningful discussion of the
cells' origin. For more information, readers can consult any of the many
good textbooks of cell biology that are now available.

THE NUCLEO-CYTOPLASMIC SPLIT

Perhaps the most important property distinguishing eukaryotic from prokaryotic cells is the division of the former into two structurally and functionally distinct parts, a highly organized pulp, called *cytoplasm*, and a
central kernel, the *nucleus*. Whereas the cytoplasm is the site of the vast
majority of metabolic processes, including protein synthesis, the nucleus

centralizes the bulk of genetic operations. The nucleus is notably where genes are stored and transcribed into messenger RNA molecules and where these molecules are further processed. In addition, the nucleus is the site of DNA replication when cells prepare to divide, and it subsequently participates in a complex set of processes leading to the formation of two nuclei.

This characteristic partition of functions does not exist in prokaryotes, in which ribosomes are commonly seen in the process of translating into proteins genetic messages that are still being transcribed from chromosomal DNA. It all takes place in a single phase. In contrast, in eukaryotes, transcription and translation are topographically separated.

SIZE

Eukaryotic cells are much larger than prokaryotes. Their diameter is usually on the order of 20 to 30 thousandths of a millimeter, with exceptions that are sometimes large enough to be visible to the naked eye. This means that a typical eukaryotic cell has a volume equivalent to that of some ten thousand prokaryotic cells. Relatively speaking, it is enormous.

OUTER COVERING

Eukaryotic cells surround themselves with a great diversity of outer coverings, which, in unicellular organisms, may vary from a thin slimy coat to structures of enormous complexity (remember the diatom mentioned in the beginning of this chapter). Pluricellular organisms all rely for their architectural support on a complex tangle of extracellular structures assembled from materials secreted by the cells. The cells themselves are just soft blobs. Deprived of their extracellular scaffoldings, trees, whales, humans, and the rest of visible life would be no more than shapeless masses of gooey stuff.

Chemically, the outer coverings of eukaryotic cells are made of a wide variety of substances, mostly proteins, carbohydrate polymers, or combinations of both. In a number of cases, as in diatoms, bones, or mollusc shells, the structures are reinforced by minerals. Interestingly, no eukaryotic cell has an outer covering made of murein or pseudomurein, the substances that make up the cell walls of eubacteria and archaebacteria, respectively. There seems to be a distinct gap at this level between eukaryotes and prokaryotes.

PLASMA MEMBRANE

Eukaryotic cells are, like their prokaryotic relatives, surrounded by a typical membrane, called *plasma membrane*. Contrary to what happens in prokaryotes, where the cell membrane may harbor important metabolic processes, such as, for example, ATP-generating oxidations, eukaryotic plasma membranes are almost exclusively specialized to serve as actively discriminating boundaries. They are richly fitted with *transporters* and *receptors* involved in exchanges and communications with the outside world (see Chapter 6). Similar phenomena take place in prokaryotes, but they are much more numerous and diversified in eukaryotes, as befits the greater complexity of the cells.

CYTOSOL

The cytoplasm of eukaryotic cells is occupied by a thick, semi-fluid material, termed *cell sap* or *cytosol*, closely related to the milieu that fills bacterial cells, except that it is occupied, in addition, by many structural components that do not exist in prokaryotes (see below). Like the inner medium of prokaryotes, the eukaryotic cytosol houses numerous metabolic systems of major importance. It is a concentrate of protein enzymes, coenzymes, and chemical intermediates involved in a variety of syntheses, breakdown processes, and transforming reactions, which may number up to several thousands. Interestingly, most of these reactions occur without the participation of oxygen. As will be seen, the utilization of this gas is largely the prerogative of special particulate entities embedded in the cytosol. This separation between anaerobic and aerobic processes is an important clue to our understanding of eukaryotic history.

The cytosol also contains, like its bacterial counterpart, the bulk of the *ribosomes*. These particles are made in the nucleus, and they also receive from the nucleus the messenger RNA molecules that provide them with the blueprints of the protein molecules they assemble. Protein synthesis itself takes place in the cytosol, from which the proteins are directed toward their final location within the cells by special *targeting* mechanisms (see below).

CYTOMEMBRANE SYSTEM

All eukaryotic cells contain a number of membrane-enclosed compartments of varying shapes and contents, organized into several distinct subsystems endowed with characteristic structural and functional specializations and forming together what is known as the *cytomembrane system*.[2] Each of these compartments is completely surrounded by a membranous envelope. They are nevertheless able to share and exchange contents by means of transient intermembrane connections, so that materials can circulate from compartment to compartment by *vesicular transport*, always separated from the cytosol by a closed membrane.

Processes also exist whereby new intracellular vesicles may detach from invaginations (infoldings) of the plasma membrane, enclosing within their insides whatever extracellular materials have been trapped in the invagination. This mechanism, called *endocytosis* (Greek for "into the cell"), introduces extracellular materials into the cytomembrane system. Conversely, existing vesicles may join with the plasma membrane, thereby discharging their contents outside the cells, while their envelope is added to the plasma membrane. Known as *exocytosis* (Greek for "out of the cell"), this process allows materials present within the cytomembrane system to be discharged out of the cell.

Thanks to the transient connections that link its various compartments with each other and with the extracellular milieu, the cytomembrane system is the site of a dual traffic concerned with what may be designated import-export. In the import direction,[3] all sorts of complex molecules and objects are engulfed from the outside by endocytosis and introduced into the system, most often to be broken down in specialized pockets, called *lysosomes* (from the Greek *lyein*, to dissolve, and *sôma*, body), within which the engulfed materials are exposed to an acid juice containing powerful digestive enzymes capable of fragmenting all major biological constituents, as do the digestive juices of our stomach and intestinal tract. The products of this chemical breakage are able, thanks to their small molecular size, to traverse the lysosomal membrane and reach the cytosol, where they enter general metabolism. Initially serving for the capture and utilization of food, this mechanism has been adapted to a variety of functions, including, as will be seen in the next chapter, the fight against pathogenic bacteria.

The export part of the system[4] serves in the manufacture and discharge of secretion products. These are mostly proteins made by *membrane-bound ribosomes* (see protein targeting, below) that inject their products directly into a cavity of the system. After their synthesis, the proteins travel through various specialized parts of the cytomembrane system, undergoing a number of chemical changes, as well as additions, notably of carbohydrate components, to be finally unloaded outside the cell by ex-ocytosis. The export system also makes the enzymes involved in intra-cellular digestion. But these, thanks to a special targeting mechanism, are discharged into the lysosomes instead of into the outside medium.

CYTOSKELETON

Another characteristic feature of eukaryotic cells is the presence of various intracellular structural elements that prop up the cells and consolidate them internally, preventing them from collapsing under their own weight and helping them to conserve their shape. Grouped under the general name of cytoskeleton, these elements are made of special protein molecules.

In most instances, the proteins involved conserve the linear confor-mation of their peptide chains and join, like the strands of a string, to form various *filaments*, which are sometimes twisted, woven, or inter-linked into three-dimensional structures of diverse shapes. In two par-ticularly important groups of structural elements, the protein chains are, like most proteins, folded into globular units. These spontaneously as-semble into elongated structures thanks to special mutually binding sites—of the mortise-tenon kind (see Chapter 1)—with which they are equipped. *Actin fibers*[5] and *microtubules*[6] are the two main structures formed in this way.

MOTOR SYSTEMS

Often associated with the cytoskeleton are motor elements that allow the cells to shift their inner parts with respect to each other and, in some cases, to move around as whole cells. The active components of these systems are special transducing proteins that convert to mechanical work the energy released by the splitting of ATP.[7]

Such associations are sometimes organized, with a number of addi-tional proteins, into elaborate propulsion organelles. Among these, a par-ticularly important group is represented by *cilia* and *flagella*, which, from protists to higher animals, are involved in a wide variety of cell move-ments. The cilia are short and act by beating. The flagella are long and have an undulating movement. Both organelles depend on the same char-acteristic structure, a highly complex arrangement of microtubules and proteins.[8] Note that many bacteria, spirochetes, for example (see Chapter 8), are also equipped with flagella, but these are totally different from eukaryotic flagella. Instead of undulating, bacterial flagella rotate, at the astonishing speed of several thousand revolutions per minute. Their shafts are rigid, corkscrew-shaped structures made of proteins that bear no relationship to the proteins of eukaryotic flagella.

Another machinery made of microtubules and motor systems is the *mitotic spindle*, the structure involved in the separation of daughter chromosomes in cell division, as will be described later.

Animal locomotion depends on molecular arrangements very different from those involving microtubules. *Myofibrils*, the functional units of animal muscles, have actin as their cytoskeletal element and operate by shortening (contraction).[9]

ORGANELLES OF OXYGEN METABOLISM

The cytoplasm of most eukaryotic cells contains a variable number—up to several thousand in some cases—of discrete, particle-shaped, membrane-bounded *organelles* (small organs) typically connected with oxygen metabolism. Several distinct types of such organelles are known.

Peroxisomes These are small granules, about one half of one thousandth of a millimeter in diameter, bounded by a single membrane, often described in the morphological literature by the name of *microbodies*. Present in one form or another in the vast majority of eukaryotic cells, peroxisomes are involved in a number of oxidative reactions in which oxygen is utilized by way of hydrogen peroxide—hence their name—and without the coupled assembly of ATP. Their role in the supply of energy is therefore limited. They nevertheless accomplish important functions in the metabolism of a number of substances, prominently lipids, as revealed by human pathology.[10] Some peroxisomes are known under a different name recalling a special metabolic property.[11]

Mitochondria These organelles, likewise utilizing oxygen and almost universally distributed among eukaryotic cells, are particles about the size of bacteria—we shall see that this is more than a fortuitous coincidence—surrounded by two membranes of which the inner one expands into numerous foldings, or *cristae* (ridges). Their shapes vary from almost spherical to slenderly filamentous, explaining their name, which comes from the Greek *mitos* (filament) and *chondros* (grain). Often referred to as the cells' "power houses" for this reason, mitochondria are, in the whole eukaryotic world, the central sites of oxidative phosphorylations, that is, cellular oxidations coupled with the assembly of ATP. The systems involved in these all-important processes are composed of a complex set of electron carriers,[12] or *respiratory chain*, embedded in the inner mitochondrial membrane.

Hydrogenosomes These are membrane-bounded organelles, present only in a few selected protists and fungi occupying habitats that are poor in oxygen but not necessarily totally devoid of this gas. Probably related to mitochondria (see next chapter), hydrogenosomes have as a remarkable property, which accounts for their name, the ability to produce molecular hydrogen in the absence of oxygen. They can also utilize oxygen.

Chloroplasts This name, derived from the Greek *chlôros* (green) and *plastos* (fashioned), designates the characteristic photosynthetic organelles of unicellular algae and green plant cells. Larger than mitochondria and similarly surrounded by two membranes, chloroplasts are filled with stacks of flattened membranous sacs, the *thylakoids*, which contain the light-utilizing systems. It will be remembered that the reactions catalyzed by these systems are associated with the production of molecular oxygen, as are those that occur in cyanobacteria (see preceding chapter). The name *plastids* is given to colorless chloroplast relatives present in plant cells that do not carry out photosynthetic processes.

NUCLEUS

The eukaryotic nucleus is separated from the cytoplasm by a *double-membranous envelope* belonging to the cytomembrane system, reinforced by an inner framework, the *lamina*, made of cytoskeletal elements. This assemblage is pierced with *pores* that mediate all the exchanges between nucleus and cytoplasm. The genetic material is confined within this envelope, organized with proteins into a number of discrete entities, the chromosomes. The nucleus is basically an RNA factory. In it, all the different kinds of RNAs that take part in protein synthesis and other functions are synthesized by transcription of the corresponding DNA genes and further processed by splicing and other modifications. With the exception of a few RNA molecules involved locally in this processing, all these RNAs are conveyed to the cytoplasm, where they carry out their various functions. This transport takes place through the pores of the nuclear envelope, with the help of special protein carriers that are allowed inside in naked state and return to the cytoplasm in combination with the RNAs.

A special intranuclear structure, the *nucleolus*, is the site of a particularly important activity, the manufacture of ribosomes, which, being relatively short-lived, have to be continually provided to the cytoplasm in large quantities. The nucleolus houses the production and processing of ribosomal RNAs and their association with proteins imported from the cytoplasm. The ribosomes thus formed are delivered into the cytoplasm, ready to carry out protein synthesis. The nucleus itself does not make proteins and is virtually devoid of metabolic and energy-yielding systems. All the NTP building blocks used for the synthesis of RNA (and DNA, see below) come from the cytoplasm through the nuclear pores. All nucleo-cytoplasmic exchanges take place through these openings. This complex, two-way molecular traffic is strictly controlled by highly elaborate mechanisms.

MITOTIC DIVISION

Contrary to prokaryotes, which tend to multiply exponentially whenever conditions allow, eukaryotic cells can remain without dividing for a long time, even indefinitely, subject to complex controls regulating what is known as the *cell cycle*. Special triggers—of burning interest to all those who study cancer, which is essentially due to unchecked cell multiplication—awaken the cells from this stationary phase and stimulate DNA replication in the nucleus, leading to doubling of the chromosomes. The nuclear envelope subsequently breaks into pieces and is replaced by an impressive scaffolding, called the *spindle* because of its shape. This structure, made largely of microtubules combined with special motor elements (see above), acts mechanically to separate the two chromosome sets and drag one set toward one spindle pole and the other to the opposite pole. There, a remarkable self-assembly process surrounds each chromosome set by a newly formed envelope, with, as a result, the genesis of two identical nuclei. This mode of cell division is called *mitotic division*, or *mitosis*.

PROTEIN TARGETING

With very rare exceptions,[13] all the proteins of a eukaryotic cell are made in the cytosol and are conveyed to their final location within the cell by a variety of mechanisms that all rely on a specific interaction ("recognition") between a short amino acid sequence (*targeting sequence*) of the protein molecule and a receptor present on the target's surface. Illustratively described as postal addresses utilized by the cellular mailing systems, these targeting sequences represent an additional piece of information written into the protein sequences, which, thus, not only determine the structural and functional properties of the molecules (see Chapter 1), but also specify the molecules' location inside the cell.

Targeting sequences may work *co-translationally* or *post-translationally*, that is, while the protein is being synthesized (while the RNA message is being translated), or afterwards. In the former case, the protein-making ribosome is "pinned," so to speak, to a membrane surface by the end of the protein chain it is in the process of synthesizing. This end carries a targeting sequence that interacts with a specific membrane receptor. As a result of this interaction, the membrane rearranges locally into a tunnel through which the growing peptide chain is directly injected into the lumen of the compartment delimited by the interacting membrane. This mode of transfer is characteristically involved in the synthesis of secretory proteins (see above). Membranes with ribosomes attached in this manner have a rough appearance in cross-section.[14]

Post-translational transfer takes place in most other instances; it conveys the finished proteins to their intracellular location by means of sophisticated mechanisms that often involve the participation of special proteins known as "chaperones." Peroxisomes, mitochondria, chloroplasts, and the nucleus all receive their proteins by post-translational transfer, dependent in each case on different targeting sequences. Vesicular transport may also be directed in certain cases by targeting sequences. Lysosomal enzymes, for example, are diverted from the secretion machinery to their destination with the help of such sequences (see above).

SUMMARY

Large size, a cytosol richly endowed with enzymes involved mostly in anaerobic metabolism and, in addition, housing the protein-synthesizing ribosomes, compartmentation by membranes into multiple pockets specialized in import-export exchanges with the external milieu, internal shoring by cytoskeletal structures often associated with motor elements, cytoplasmic organelles involved in the utilization and (only in photosynthetic cells) production of oxygen, organization of the genetic material into structured chromosomes confined within an envelope of both membranous and cytoskeletal nature, strict separation between gene transcription (nucleus) and expression (cytoplasm), mitotic cell division subject to elaborate control: such are the essential characteristics of virtually all eukaryotic cells, be they unicellular organisms or the components of plants, fungi, or animals. No doubt, these characteristics were all present already in the common ancestor of all eukaryotic cells. The genesis of these characteristics from some prokaryote ancestor is what needs to be explained.

Faced with the problem posed in this way, one may well wonder whether its solution will ever be found or, even, whether a "natural" solution exists. Could not the number and complexity of the mutually complementary innovations that have to be explained exemplify the "irreducible complexity" claimed by some to demand the intervention of "something else?"[15] Fortunately, we are nowhere near such a stage. As I shall show in the next chapter, modern biology has provided a number of revealing clues that illuminate certain aspects of the problem. Possibly the most significant of these concerns the role of oxygen in the genesis of eukaryotes, a role so important that it warrants special treatment now.

THE GREAT OXYGEN DIVIDE

We have seen in the preceding chapter how the development of oxygen-generating photosynthesis must have upset the physical conditions to which all early living forms were adapted, creating one of the most stringent and devastating bottlenecks in the whole history of life. There are good reasons to believe that the genesis of eukaryotes was also drastically influenced by the rise of oxygen in the atmosphere. To begin with, there is the matter of timing.

TWO CRITICAL DATES

Two important landmarks flank the history of eukaryotes. On one side are Woese's sequencing results[16] indicating that the line leading to eukaryotes separated from the two prokaryotic lines *very early* after the tree of life first branched out from its root, that is, if the evidence in Chapter 3 is to be trusted,[17] no later than some 3.5—perhaps even 3.8—billion years ago. On the other side, there is strong evidence that the common ancestor of eukaryotes must have been aerobic, as it almost certainly contained mitochondria as well as peroxisomes. If, as geochemical findings seem to indicate, the level of atmospheric oxygen was too low to support aerobic organisms before about 2.3 billion years ago, the common ancestor of eukaryotes cannot have lived earlier than that date.

There thus seems to be, between what may be called the "founder" of the eukaryotic line and the common ancestor of all present-day eukaryotic life, a gap of at least 1.2 billion years, not counting the time it took for oxygen to rise to a level allowing aerobic life. This estimate may need revision in view of the uncertainties surrounding the time of the first appearance of life on Earth. Doubts have also been expressed with respect to the validity of sequencing phylogenies extended, as done by Woese, to extremely ancient times. But the gap is certain to remain huge, at least several hundred million years. During this immense amount of time, one of the most extraordinary and momentous developments in the history of life took place: the transformation of what was most likely a primitive bacterium (see Chapter 3) into a much more complex cell type that, in turn, was to give rise to the rich array of protists and to the whole visible panoply of plants, fungi, and animals, including ourselves. Had this transformation not taken place, the living world of today would still contain only bacteria.

The time limits just mentioned do not necessarily mean that the transformation of a prokaryotic into a eukaryotic cell took the whole of the time allowed by the record, only that this amount of time was available for the transformation. Nor do these limits imply that the transformation followed a single line, determined, as is often assumed, by a number of improbable "quantum jumps" that guarantee its unique character. It is quite possible—I would even say probable—that the ancestral eukaryote emerged from an evolutionary bottleneck preceded by a long, complex history. I shall come back several times to this important question.

THE OXYGEN CONNECTION

Whatever the events that took place during the gap mentioned above, they must, if the evidence referred to is correct, have occurred within the framework of anaerobic life. It is striking in this respect that all the main features of eukaryotic cells that have just been surveyed are, with the exception of the organelles of oxygen metabolism, associated in major part with biochemical mechanisms that do not involve oxygen. The rare exceptions to this generalization—biochemists may think, for example, of the oxygenations and hydroxylations catalyzed by components of the cytomembrane system—could have been late additions. In view of these facts, it is tempting to assume that the main eukaryotic features were acquired before oxygen started rising in the atmosphere.

Strong support for this hypothesis comes from a discovery of capital importance that has projected a most revealing light on the genesis of eukaryotic cells. It is now established beyond reasonable doubt that at least two major organelles of oxygen metabolism, namely mitochondria and chloroplasts, were once free-living bacteria, which were adopted at a remote time by host cells within which they became integrated as *endosymbionts*, a Greek-derived term that literally means: living (*biont*) together (*sym*) within (*endo*).[18]

I shall come back at length to this epoch-making event. For the time being, let me simply emphasize an important implication of the endosymbiotic origin of the organelles concerned. It means that the actual development of these organelles was *not part of the eukaryotic transformation*. It belongs to prokaryotic history, which still includes surviving descendants of the transformation's actors, as will be mentioned later. The organelles were acquired *ready-made* and only their subsequent integration within the host cells' economy is part of eukaryotic history.

The origin of peroxisomes is unfortunately not known. It is possible that these organelles also originate from endosymbiotic bacteria. But this cannot be affirmed, as the evidence available so far on this topic is ambiguous. Whatever their origin, peroxisomes must obviously go back, like the mitochondria and the chloroplasts, to a time in the history of life, when oxygen was already present in significant amounts in the atmosphere, that is, less than 2.3 billion years ago.

A TWO-PHASE MODEL OF EUKARYOTE GENESIS

The apparent division of eukaryotic cells into an anaerobic and an aerobic part suggests a development in two phases. The first, anaerobic phase is assumed to have taken place before the rise in atmospheric oxygen and to have led to the development of anaerobic, heterotrophic cells possessing all the main properties of eukaryotic cells, with the exception of oxygen-linked organelles. This phase may have taken a greater or lesser part of the 1.2 billion-year stretch left open for it. Quite possibly, a wide variety of such cells may have existed, subsisting largely on bacteria.

The second phase of this hypothetical scenario is pictured as being

initiated by the emergence of oxygen in the atmosphere. As we saw in
the preceding chapter, this event may have signalled a widespread extinction of living forms, sometimes called the oxygen holocaust. It is quite
possible that most of the primitive eukaryotes supposed to have arisen
in the first phase fell victim to this catastrophe, leaving only those that
had acquired the necessary defenses (or had found refuge in an oxygen-
free niche). In the end, only one form would have made it safely through
the bottleneck and survived the resulting competition, thus accounting
for the single ancestry of extant eukaryotes (see above).

Surprisingly, few theories proposed for the origin of eukaryotes postulate a long anaerobic phase, followed by endosymbiont adoption. In
popular accounts, the establishment of an endosymbiotic relationship is
often presented as the starting point of eukaryote development, or even
its triggering event. All is taken to have started with a "fateful encounter"
between two different bacteria that established some kind of mutually
beneficial association in which, eventually, one became dominant and the
other submissive.

If this is what happened, then we are faced with the question as to
what took place during the huge span of time that has elapsed between
the first branching out of the eukaryotic line and the adoption of endosymbionts. Another difficulty is that the alternative theories that are
proposed usually offer no explanation for the acquisition of the many
characteristic components of eukaryotic cells other than the organelles of
oxygen metabolism. The genesis of these features is rarely addressed by
the defenders of the fateful encounter model. Finally, the model throws
little light on the manner in which the encounter ended with the en-
slavement of one participant by the other.

It could be argued that the above objections rest on questionable data.
The methods of dating by molecular sequencing results, for example, are
the object of vigorous debates. There is also some disagreement on the
actual level of atmospheric oxygen before the critical date of 2.3 billion
years ago. Some authors believe this level to have been high enough to
support some forms of aerobic life. This book hardly lends itself to a
detailed discussion of these highly specialized issues. What I believe to
be important is that an alternative model, not open to the formulated
objections, can be proposed. This is the model I have adopted in this
book. Its main features will be described in the next chapter.

Note that even if the proposed model should turn out to be incorrect, the possibility that most of the eukaryotic properties were acquired before the prokaryotic ancestors of present-day organelles were adopted as endosymbionts seems sufficiently plausible, if not likely, to deserve being seriously entertained. Furthermore, whatever model is adopted, many of the mechanisms envisaged could still be relevant.

10. The Mysterious Birth of Eukaryotes

A POSSIBLE
PATHWAY

ACCORDING
to the model
outlined in
the preceding chapter,
the history of eukary-
otic cells is divided
into two parts: an anaerobic phase, in the course of which were developed
all the main eukaryotic properties, with the exception of the organelles
of oxygen metabolism, and an aerobic phase dominated by the acquisition
of those organelles.

PHASE I: BEFORE OXYGEN

We have seen in Chapter 8 that the eukaryotic line probably detached
from the two prokaryotic lines early after the tree of life first emerged
from its root. By what kind of cell was this process inaugurated? This is
a much-debated question.

THE FOUNDER CELL

Early sequencing results first suggested an archaebacterial origin of the
eukaryotic line. Later, however, a number of eukaryotic genes were found
to be more closely related to eubacterial than to archaebacterial genes. In
the opinion of many investigators, this genetic blend could be the outcome of horizontal gene transfers, such as are believed to have occurred
on a large scale in the early days of life (see Chapter 8).

Some workers, however, point out that the mixture is apparently not random. To put a complicated matter in simple terms, the eubacterial genes mostly code for "housekeeping" enzymes, whereas the archaebacterial genes tend to define components of genetic information-transfer systems. Or, put in even more simplistic terms, the cytosol of the ancestral cell appears to be of eubacterial origin, its nucleus of archaebacterial origin. On the strength of this dichotomy, the suggestion has been made that the eukaryotic line was initiated by the fusion (the first of a number of fateful encounters) between a eubacterium, which ended up providing most of the cytoplasm to the chimera, and an archaebacterium, which furnished the chimera's genetic machinery. Other models have also been proposed, including that of an endosymbiotic origin of the eukaryotic nucleus.

Leaving this question to the experts, I shall take as a working hypothesis that the founder of the eukaryotic line, whatever its origin, was a relatively simple, anaerobic, heterotrophic (subsisting on organic food) organism resembling present-day prokaryotes in its main properties. I shall further assume that this prokaryote did not have a cell wall. This hypothesis is derived from the fact, noted in the preceding chapter, that no eukaryotic cell is known that possesses a covering made of the same chemical substances as bacterial cell walls. This fact suggests that the ability to build a typical prokaryotic cell wall was lost in the eukaryotic line.[1] It is possible, as will be mentioned later, that this loss may have played a significant role in the prokaryote-eukaryote transition.

THE PATHWAY FOLLOWED

Barring evidence to the contrary, it seems reasonable to assume that the ancestral prokaryote converted into a primitive eukaryote by a long succession of largely conserved intermediates each of which was only just a little less prokaryotic and a little more eukaryotic than its predecessor.

A first modification, which can be safely postulated in the framework of this hypothesis, is that the cells slowly grew bigger. They did not fundamentally change; they went on using the same enzymes and following the same metabolic pathways as before; they just made more of everything per cell, simply by delaying DNA replication and the onset of division. Interestingly, something of the kind happens to bacterial cells that have been stripped of their external wall (by an enzyme called lysozyme) or prevented from building a wall (by penicillin) in a medium where they are protected against osmotic bursting. The resulting naked cells, known as protoplasts, increase in size.

A second predictable modification is that the cell membrane expanded, perhaps even overexpanded, making numerous convolutions around an increasingly contorted cell body. Such a change, which, incidentally, may have been facilitated by the loss of a rigid external wall, was an almost obligatory concomitant of cellular enlargement, as more surface area was needed to allow the increasing exchanges of matter with the environment required to support the growing cell mass.[2]

The next step, which is seen as the crucial event in the proposed scenario, is assumed to be the occasional formation of closed, membrane-bounded, intracellular vesicles, splitting off from deepening invaginations of the cell membrane. Such phenomena, which, initially, could have depended on no more than the natural self-sealing property of lipid bilayers (see Chapter 6) helped by surface tension, are taken to have initiated the genesis of the cytomembrane system, leading, in the course of a protracted and complex evolutionary history, which also involved cytoskeletal and motor elements (see below), to a progressive differentiation of the vesicles into distinct parts, functionally specialized in a variety of import-export exchanges with the outside, such as are known today.

There is much to be said for this model. Membranes, we have seen, always arise from membranes—remember Blobel's aphorism, *omnis membrana e membrana* (see Chapter 2). It is therefore likely that the inner membranes of eukaryotes arose from the surrounding membrane of their prokaryotic ancestor. In agreement with this hypothesis, the membranes of the two cell types have in common a number of genetically related functional systems. Most impressive are the systems involved in secretion. We have seen that, in eukaryotic cells, the ribosomes that synthesize secretory proteins stick to certain membranes of the cytomembrane system (rough endoplasmic reticulum[3]) and directly inject the proteins they manufacture into the cavities bounded by these membranes. In certain bacteria, ribosomes attached to the inner face of the cell membrane similarly deliver secretory proteins into the extracellular medium. Remarkably, the two systems depend on closely similar targeting sequences for directing the ribosomes to their membrane receptors, to the point of being able to obey each other's signals. This similarity is clear evidence of an evolutionary kinship between the two systems and strongly supports the view that intracellular vesicles studded with ribosomes originally arose from similarly studded invaginations of the cell membrane, eventually to become what is now known as the rough endoplasmic reticulum (see below).

Another development, which, like membrane expansion, was to some extent mandated by the growing cell size, was the acquisition of cytoskeletal and motor elements. Little is known concerning the origin of these elements, which, as we have seen, are made essentially of certain specific proteins. Attempts at identifying in bacteria genes possibly ancestral to the eukaryotic genes coding for these proteins have gleaned limited results so far.[4] It may be that fairly radical genetic innovations lie behind the eukaryotic development. It could also be, however, that the relevant innovations were fairly commonplace but of little use to prokaryotes and therefore not retained by natural selection.

The biologist Lynn Margulis, known for her early defense of, against considerable opposition, the endosymbiont theory, has proposed that eukaryotic flagella and cilia, motor organelles that she groups under the name "undulipodia," are also derived from endosymbiotic bacteria, which she believes to be related to present-day spirochetes.[5] This, according to her hypothesis, is how eukaryotic cells acquired microtubules, which are the main constituents of the motor organelles but have many other functions as well. The evidence put forward in support of this theory has, however, failed to convince the majority of experts. As we have seen in the preceding chapter, there is no evidence of either structural or functional kinship between eukaryotic flagella and the motor appendages of spirochetes.

It seems more likely that tubulins and the resulting microtubules arose first, in relation with one of the various functions carried out by these entities in present-day cells (mitotic division?) and that the characteristic structures of eukaryotic flagella and cilia, which share a highly complex arrangement of microtubules and associated proteins,[6] developed later, though perhaps early enough for the common ancestor of eukaryotes to be motile, propelled, as are a number of protists, by beating cilia or undulating flagella.

Interestingly, the characteristic segregation of the genetic machinery within a central nucleus, which is the hallmark of eukaryotes, could possibly have been initiated by the membrane expansion process to have led to the cytomembrane system. In prokaryotes, the chromosome is anchored to the cell membrane. Invagination of this part of the cell membrane, in the general framework of cytomembrane development, would have dragged the chromosome into the interior of the cell. The resulting vesicle, perhaps with the participation of other vesicles, could conceivably have folded around the attached chromosome and enclosed it within a double-membranous envelope, related to the cytomembrane system as is the nuclear envelope in present-day eukaryotes. Cytoskeletal elements would then have been added later to reinforce this structure. This, incidentally, is how new nuclear envelopes form, at the end of mitosis, from parts of the cytomembrane system that fuse together around the segregated chromosomes.

Thus, the basic separation between nucleus and cytoplasm could have been initiated as part of the formation of the cytomembrane system. After that, a large number of innovations, about which virtually nothing is known, must have taken place to produce the highly complex organization of eukaryotic nuclei and the elaborate machinery involved in mitotic division.

In summary, the pathway proposed for the prokaryote-eukaryote transition is centered on a process of *membrane expansion and vesiculation*, associated with *cellular enlargement* and supported by the coevolutionary development of *cytoskeletal and motor systems* of increasing complexity. A process of this sort could, indeed, as postulated earlier, have involved a very large number of viable intermediates with progressively modified properties. Furthermore, plausible mechanisms can be suggested for some of its main steps. But for these to take place and for their acquisitions to become genetically transmissible, they must have enhanced the cells' ability to survive and produce progeny under the prevailing environmental conditions. What could those advantages have been?

INTRACELLULAR DIGESTION, MOTOR OF THE EUKARYOTIC BREAKTHROUGH?

Imagine a heterotrophic bacterium of the kind that could have been ancestral to eukaryotic cells. It subsists on external food, which, like all present-day heterotrophic bacteria, it digests with the help of secreted *exoenzymes*. For obvious reasons, such an organism is forced to reside inside its food supply or, at least, in close juxtaposition with it, so that the secreted enzymes can accomplish their digestive functions without being immediately lost in the surroundings.

Now visualize such a cell beginning to undergo the membrane expansion and internalization process postulated in our model. All extracellular materials passively dragged into an invagination of the cell membrane will become segregated inside the resulting vesicle, as in endocytosis. On the other hand, exoenzymes that were secreted outside the cell by the piece of membrane involved in vesiculation will henceforth be discharged into the vesicle, where they will be able to act on the trapped material, as occurs in lysosomes. *Extracellular digestion has become intracellular digestion.*

The benefit of this transition, even in its primitive and random form, is enormous. Thanks to a seemingly trivial modification, the cell concerned and any of its progeny endowed with the same vesiculation ability have ceased to depend for their survival on intimate contact with a nutritive substratum. They are now free to roam around, to invade ponds and oceans, subsisting on food captured by infoldings of their membrane and digested within the resulting intracellular vesicles. This ability clearly entailed an invaluable evolutionary asset—I have referred to it as the beginning of cellular *emancipation*—so that any genetic modification likely to favor membrane expansion and the associated folding and fusion phenomena would have been strongly advantaged by natural selection.

Note that the first primitive vesicles proposed in this hypothesis combine the properties of three major components of the cytomembrane system: *endocytic vesicles*, containing material captured from the outside; *lysosomes*, sites of the digestion of this material; and parts of the *rough endoplasmic reticulum*, providers of the necessary digestive enzymes. The subsequent evolutionary history of the system may be readily pictured as a progressive separation of these functions into distinct, differentiated parts, finally leading to the complex cytomembrane system of extant cells. As we have seen, cytoskeletal and motor elements presumably participated in this evolutionary process.

THE FIRST EUKARYOTES

According to the model sketched earlier, which resembles in many respects that long defended by the British biologist Thomas Cavalier-Smith, the first eukaryotes were large, anaerobic, heterotrophic, possibly motile, unicellular organisms with, at least in primitive form, all the main features of extant eukaryotic cells except cytoplasmic, oxygen-related organelles. The organisms caught their food by active endocytosis and digested it intracellularly within lysosomes. For all we know, such organisms could have arisen long before the 2.3-billion-year limit set by the oxygen divide; and they could have covered Earth with a wealth of thriving varieties, of which all but one disappeared without leaving any fossil trace or lasting progeny.

Some years ago, investigators thought they had found descendants of ancient witnesses to this history. Two groups of unicellular organisms devoid of mitochondria, diplomonads and microsporidia, were found by sequencing to go back to particularly remote times. The organisms possess all the other characteristics of eukaryotic cells; they thus looked for all the world like descendants of a line that had detached from the hypothetical primitive eukaryotes before the acquisition of endosymbiotic organelles. Alas! The same techniques of molecular sequencing have dashed the hopes they had raised. Not only has the great antiquity of the organisms been questioned by some investigators, but genes of mitochondrial origin have been identified in their nuclei. If they lack mitochondria, this is apparently not because they never had such organelles but because they have lost them. At present, no eukaryotic cell derived from an ancestor that never possessed mitochondria (or mitochondria-related organelles, such as hydrogenosomes, see below) is known.

This, however, hardly justifies the conclusion, sometimes drawn, that primitive eukaryotes of the type envisaged never existed. Absence of evidence is not evidence of absence. It is an incontrovertible fact that the main characteristics of eukaryotic cells *were* acquired at some stage in the history of life. How this happened is not known. The hypothetical scenario proposed above for this transformation may be totally wrong. But, at least, it is consistent with the meager clues available and has the merit of providing a possible selective driving force for the process. The truth is that no alternative model has yet been put forward. As to the timing of the transformation, there is a strong reason for putting it before rather than after the acquisition of endosymbionts: the transformation offers the most likely mechanism by which this phenomenon could have occurred.

Phase 2: After Oxygen

As we saw in the preceding chapter, endosymbiont adoption is often visualized as the outcome, largely unexplained, of a fateful encounter between two bacterial species. However, if there is any truth in my proposed model, the encounter was not between two bacteria, but between a bacterium and a primitive eukaryote of the kind just described. In fact, it was *not an encounter* in the usual sense of the word, but rather the *capture* of a passive victim by an active hunter. How such events may be pictured will be briefly considered in the following pages.

THE ENDOSYMBIONTS

The main organelles known to be derived from endosymbiotic bacteria are the *mitochondria*, which are the sites of the principal oxidation reactions linked to the assembly of ATP (oxidative phosphorylations), and the *chloroplasts*, which are the agents of photosynthesis in unicellular algae and plants.

Among the many pieces of evidence supporting the bacterial origin of these organelles, the most convincing is the presence, in mitochondria and chloroplasts, of still-functional vestiges of an ancient genetic apparatus of prokaryotic character. This apparatus consists of a small number of genes, rarely exceeding a few tens, and all that is needed to replicate, transcribe, and translate these genes. Other clues include metabolic similarities and genetic kinships between the organelles and certain extant bacteria.

In the opinion of the vast majority of investigators, these proofs are conclusive. Even the organelles' bacterial ancestors or, to be more precise, their closest relatives among present-day bacteria have been identified. They are, for the mitochondria, aerobic organisms known under the name of *α-proteobacteria* and, for the chloroplasts, *cyanobacteria*, those photosynthetic bacteria believed to be responsible for the first generation of atmospheric oxygen (see Chapter 8).

Possibly also derived from endosymbionts—but this is far from certain—are the *peroxisomes*, which accomplish oxidative reactions of primitive character that, contrary to those that occur in mitochondria, are not coupled to the assembly of ATP. The origin of peroxisomes is not known. It has been suggested that they also may have arisen from endosymbiotic bacteria. This possibility is consistent with the metabolic activities of the organelles. Added together from the properties of all known members of the peroxisome family,[7] these activities cover a very wide range, as would be expected of an autonomous ancestral organism. However, peroxisomes contain no trace of a genetic system, and the molecular data obtained so far are ambiguous.

The common ancestor of eukaryotes almost certainly contained both mitochondria and peroxisomes. The two kinds of organelles are present, in one form or another, in the vast majority of present-day eukaryotic cells. The rare exceptions, of which several are already known to be the outcome of evolutionary regressions, do not suffice to invalidate the generalization. As to the chloroplasts, they were probably not present in the common ancestor but were acquired later in the branch leading to the photosynthetic eukaryotes. The inverse hypothesis, a loss in the non-photosynthetic branches, seems less plausible. It is noteworthy, in this connection, that chloroplasts generally contain a larger number of genes than mitochondria, possibly indicating a more recent adoption.

CAPTURE OF THE ENDOSYMBIONT ANCESTORS

Granted the nature of the postulated host cells, it seems most likely that the endosymbiont ancestors were originally caught by *phagocytosis* (from the Greek *phagein*, to eat, and *kytos*, cell), a term coined by the Russian-French zoologist and immunologist Ilya Metchnikoff, famous for the discovery that white blood cells protect against bacterial infections by engulfing the disease-causing bacteria and destroying them intracellularly. As already noted by Metchnikoff, this function is just a particular specialization of a more general process used by heterotrophic protists—and by our hypothetical primitive eukaryotes, which, for this reason, are sometimes referred to as *primitive phagocytes*—for the capture of food. As we have seen, the term endocytosis, actually derived from phagocytosis,[8] now designates the more general process, which, from its original relationship to food uptake, has become adapted in higher eukaryotes to a wide variety of functions. Lysosomes are the sites in which the captured materials are digested (and where the caught bacteria are killed and broken down).

Rare cases are known in which phagocytic capture is not followed by the death and digestion of the engulfed bacteria. The agents of tuberculosis and leprosy, for example, are not killed by the cells that catch them, but, instead, settle within those cells and proliferate. The colonized cells react to this proliferation by growing into giant cells, characteristic of the diseases, but eventually succumb. Exceptionally, the eating cells and their prey both survive and establish a relationship of mutual tolerance. It may happen that the partners of such associations, having lost some essential property, become dependent on each other for their survival. The relationship then becomes authentically symbiotic. Although rare, such phenomena are sufficiently frequent to have prompted the creation of a new discipline, *endocytobiology*, whose specialists meet regularly to compare their findings.

Added to the fact that the endocytic way of taking up extracellular material is a general eukaryotic property, which is not shared by any prokaryote and therefore must have developed in the course of the prokaryote-eukaryote transformation, all this evidence builds a compelling case in favor of the view that the ancestors of the organelles were indeed *caught by phagocytosis*, as surmised, rather than as the result of some fateful encounter between two kinds of bacteria.

INTEGRATION OF ENDOSYMBIONTS

The most striking feature of the endosymbionts, as compared to their bacterial ancestors, is that they have *lost the greater part of their genes*. Some of these genes probably turned out to be redundant, given the extensive support provided by the host cell, and just disappeared, in a kind of genetic "streamlining." A number of the genes, however, were transferred to the nucleus of the host cell, there to continue their function.

How this may have taken place is not too difficult to visualize. It no doubt happened from time to time that injured bacterial guests spilled out their DNA into the cytoplasm of the host cell. On the other hand, it is known from present-day technology (see Chapter 15) that DNA molecules introduced into the cytoplasm may occasionally enter the nucleus and become integrated within the genome, henceforth to be replicated and transcribed like the cell's own genes. Finally, plenty of time was available for experimentation, since bacterial cells with an intact genome remained present and could survive indefinitely with the help of their own copy of the transferred gene. Only after this copy had become redundant could it fall victim to the streamlining process mentioned above.

So far so good. But there is a hitch. Once integrated into the nucleus, the transferred gene behaves like a nuclear gene. It is transcribed locally, and the resulting messenger RNA is, like all messenger RNAs formed in the nucleus, delivered into the cytosol of the host cell, where it instructs ribosomes to synthesize the corresponding protein. This protein thus lands in the cytosol of the host cell, not inside the bacterial guest where it is needed. If the bacterium cannot do without the protein, the gene transfer cannot be completed until some mechanism has arisen whereby the protein (or the messenger RNA) can be transferred into its erstwhile owner.

As we now know, the transfer involves the proteins, which are directed to their site by specific mechanisms dependent on *targeting sequences*. The development of these sequences and that of the appropriate receptors and machineries on the surface of the captive bacteria probably represented the greatest challenge to endosymbiont adoption. Possible mechanisms involving the bacterium's secretory machineries have been considered but are too specialized for the present account. Let it simply be stated that gene transfer has occurred on a very large scale, to the point that only a small number of the original genes, rarely exceeding a few tens, has remained in the endosymbionts. This fact calls for several comments.

First, the finding that such massive gene transfer has happened allows the hypothesis, evoked earlier, that peroxisomes also have an endosymbiotic origin, even though they contain no trace of a genetic apparatus. If, as seems likely, peroxisomes were acquired before mitochondria, they could have lost all their genes, whereas mitochondria still have retained a few. Consistent with this possibility is the fact that chloroplasts, which were probably adopted after mitochondria and peroxisomes, have conserved a larger number of their original genes.

The occurrence of gene transfer from endosymbiont to nucleus has also provided a valuable tool for probing the past. Genetic vestiges of a vanished endosymbiont may still be left in the nucleus of a cell and may reveal the endosymbiont's erstwhile presence. This is how it was found that the ancestors of organisms lacking mitochondria did once possess such organelles.

Finally, the fact that gene transfer actually took place on such a large scale in spite of the obstacles it encountered is clear proof that this phenomenon must have been crucially important for the successful adoption of endosymbionts. Why this should be so is readily understood. Bacteria multiply much faster than eukaryotic cells. Unless the multiplication of the captured bacteria could somehow be curbed, they would inevitably overwhelm and stifle their host cells. We have seen that this happens to cells invaded by the bacterial agents of tuberculosis and leprosy, despite the fact that the cells respond to the invasion by greatly increasing their size. Removing an essential gene from the captives and transferring it to the captor's nucleus offers a particularly simple way of adapting the multiplication of captive cells to that of their captors. In the nucleus, replication of the transferred gene becomes synchronized with that of the nuclear genes, so that the captives are forced to adopt the captor's multiplication rate.

This is not the only advantage. For the captives to lose their independence and become increasingly integrated within the host cell's economy, nothing could have been more efficient than the transfer of their genes to their host cell's nucleus. It is evidently very advantageous for a host cell to have endosymbiont genes, of which there previously existed up to thousands of copies, housed in as many semi-independent entities, reduced to single copies present in the nucleus, where their replication can be coordinated with that of the host's genes and their transcription subjected to centralized controls.

These facts leave one important question unanswered. What is it that made the endosymbionts so vitally important to their host cells that all eukaryotes lacking endosymbionts seem to be extinct?

THE EVOLUTIONARY DRIVING FORCE
OF ENDOSYMBIONT ADOPTION

We have seen that, according to the latest evidence, no eukaryotic cell is known that does not have in its ancestry cells that contained mitochondria. This fact strongly suggests that mitochondria offered an enormous selective advantage, perhaps even a vitally important one, to their possessors, so that all the primitive eukaryotes that did not acquire these organelles were eliminated by natural selection. It has long been assumed that protection against oxygen toxicity made up this advantage. This explanation, which was already favored by Margulis in her early advocacy of the endosymbiont theory, is consistent with the hypothesis, evoked earlier, that oxygen poisoning wiped out all the primitive eukaryotes except those that had acquired endosymbionts.

Applied to mitochondria, however, the explanation does not hold water. Mitochondria, together with the α-proteobacteria with which they share the nearest common ancestor (see above), contain the most sophisticated oxygen-utilizing systems found in nature. True marvels of molecular organization, with an ATP yield near the maximum authorized by the laws of thermodynamics, these systems can be but the products of a very long evolution. This makes it very unlikely that the mitochondria could have saved the primitive anaerobic eukaryotes from the deadly oxygen attack. By the time the bacterial ancestors of these organelles had developed their sophisticated systems, the cells they are assumed to have saved would long have succumbed to the oxygen holocaust.

This does not necessarily invalidate the oxygen bottleneck hypothesis. But we must look for more primitive rescuers. The peroxisomes appear as excellent candidates for this function. Indeed, their properties are very much what would be expected of a primitive system of protection against the toxic gas. Their enzymes do nothing but convert oxygen and its noxious products into harmless water molecules, doing this by means of simple reactions that, unlike those that take place in mitochondria, are not coupled to the assembly of ATP. Peroxisomes or their close relatives are, like mitochondria, present in the vast majority of eukaryotic cells. It is thus perfectly possible that they were acquired before mitochondria. We have seen that the possible endosymbiotic origin of peroxisomes is at present a moot question. But this does not fundamentally change the proposed hypothesis. Even if peroxisomes were acquired in a different way, they could still have protected their owners against oxygen toxicity.

Granted this possibility, the fact remains that mitochondria must have provided a sufficiently powerful advantage to the cells that acquired them that natural selection eliminated *all* the cell types that did not enjoy this benefit, as seems to be the case. It is tempting to assume that mitochondria owed their selective value to their remarkable energetic efficiency. Peroxisomes, remember, contain no ATP-retrieval system. Their sole advantage, in terms of energy, would have been to provide the cytoplasm of their host cells with additional fuel arising from the fatty acids and other materials that only they are able to metabolize. For the actual generation of ATP, the cells endowed with peroxisomes remained entirely dependent on the coupled ATP-generating systems that support anaerobic metabolism. In such a context, the kind of oxidative machineries provided by the mitochondria represented a tremendous asset, possibly sufficient to explain why they would be retained by natural selection.

If this theory is correct, we may well ask why the acquisition of mitochondria did not drive out the more primitive peroxisomes. And, especially, why did no cell fitted only with peroxisomes survive? The answer to the first question is simple. By the time mitochondria were adopted, peroxisomes may have become indispensable because they were carrying out reactions that the newcomers could not perform, in particular in lipid metabolism, where peroxisomes are known from human pathology to accomplish vitally important functions (see Chapter 9). The fact that peroxisomes did not disappear after the adoption of mitochondria could thereby be explained.

As to the second question, the intensity of the selective pressure may provide the answer. If competition for available resources was fierce enough, only the better-equipped cells would be expected to survive. Note, however, that our knowledge of unicellular eukaryotes is still far from exhaustive. Perhaps representatives of the missing intermediates are still waiting to be found. Such a discovery would be most revealing.

As will be mentioned later, a new, startling theory, based on the production of molecular hydrogen by the ancestors of mitochondria, has been proposed to explain the adoption of these organelles. Before we consider this new theory, a brief look at the chloroplasts is in order.

We have seen that chloroplasts are derived from cyanobacteria, the oxygen-generating photosynthetic organisms believed to be responsible for the oxygen holocaust. According to all available evidence, the mechanisms involved in the uptake of these organisms and in their integration, including the massive transfer of genes to the nucleus and the development of specific protein-targeting mechanisms, must have been very similar to those involved in the adoption of mitochondria. There are good reasons to believe that the cells that did the acquisition already possessed peroxisomes and mitochondria. First, all the cell types that contain chloroplasts also contain the other two kinds of organelles. In addition, it is difficult to see how cells not properly protected against oxygen toxicity could possibly have come to harbor guests that actually produce the toxic gas.

The cells that adopted chloroplasts became the first *unicellular algae*, which, in turn, are ancestral to the pluricellular plants (see following chapter). Considered from an evolutionary point of view, the adoption of chloroplasts poses no special problem. The advantages the cells derived from their new acquisition are obvious. Henceforth freed from the obligation to find food, they housed photochemical factories that, in the presence of light, allowed them to live on water, carbon dioxide, and a few mineral salts. The benefits were immense, but not to the extent of creating a necessity. Cells devoid of chloroplasts continued to thrive, supported by their photosynthetic relatives, which became their food supply. Thus were born the main groups of unicellular eukaryotes out of which the whole visible part of the living world was to emerge.

DID THE FIRST MITOCHONDRIA MAKE HYDROGEN?

This question has been posed in recent years as a result of startling findings indicating that *hydrogenosomes*, those hydrogen-generating organelles already briefly mentioned in the preceding chapter, may be genetically related to mitochondria. The metabolic properties of these organelles hardly would have suggested such a possibility. Present in a small number of protists and fungi devoid of mitochondria, hydrogenosomes lack all the characteristic oxidative machineries of mitochondria. Their most typical property, which is absent in mitochondria, is the ability to generate molecular hydrogen anaerobically by a reaction linked to the assembly of ATP. In the presence of oxygen, this hydrogen is diverted toward the formation of water by an oxidizing system of primitive character. Thus, organisms endowed with hydrogenosomes can develop under anaerobic conditions, their usual habitat, but are also able to tolerate oxygen, if necessary, and, even, to take advantage of it. They are *facultative anaerobes*.

Hydrogenosomes do have some properties in common with mitochondria: they are surrounded by two membranes and they have been found in one case (see below) to contain a vestigial genetic machinery; especially, they share some genes with mitochondria. This is the discovery that has led to the conclusion that the two organelles have a common ancestry.

If such is the case, the question arises as to which metabolic properties characterized their common ancestor. In view of the kinship of mitochondria with α-proteobacteria, revealed by molecular sequencing data, there can be little doubt that their ancestor already possessed the sophisticated ATP-generating oxidizing systems they share with these organisms. In any case, it is hardly conceivable that mitochondria could have developed such elaborate systems independently, after their adoption as endosymbionts. On the other hand, the fact that hydrogenosomes have been found in several distantly related protists and, even, in some fungi indicates that the ability to produce molecular hydrogen must likewise be of ancient origin and may also have belonged to the putative bacterial ancestor hydrogenosomes have in common with mitochondria. Thus, the ancestor seems to have combined the main properties of both of these organelles.

One is thus faced with a strange case of evolutionary divergence. Starting from an ancestor simultaneously endowed with highly efficient oxidizing systems and with an anaerobic hydrogen-generating machinery, the vast majority of organelles would have kept only the former and lost the latter, becoming mitochondria. A small minority would have done the opposite and given rise to hydrogenosomes. None would have retained both machineries. A divergent adaptation to aerobic and anaerobic milieus could conceivably explain this occurrence, which does, however, remain puzzling.

The new findings also raise another intriguing question: which of the two properties offered the selective advantage host cells derived from adopting the ancestors of the organelles? All earlier theories have invoked the possession of oxidizing systems with a high ATP yield as the main benefit. This is what was suggested earlier. But there is now the alternative possibility that it was the *ability to produce hydrogen* that made the endosymbionts useful to their host cells.

A theory based on this second eventuality has been proposed by the discoverer of hydrogenosomes, my erstwhile collaborator and present colleague at the Rockefeller University in New York, Miklos Müller, together with an American investigator stationed in Germany, William Martin. As suggested by these workers, the host would have been an organism related to present-day *methanogens*. These microbes (see Chapter 8) are strictly anaerobic, autotrophic archaebacteria that use molecular hydrogen to convert carbon dioxide into methane by a reaction coupled to the assembly of the ATP they need to satisfy their energy requirements. According to the proposed theory, the benefit host cells derived from the endosymbionts was the hydrogen they needed as fuel for making ATP, not ATP itself.

A detailed discussion of the two competing theories does not belong in this book. Let it simply be pointed out that the model based on hydrogen supposes an encounter between two typical bacteria. Like other fateful encounter models, it does not include the participation of a primitive phagocytic host cell and says nothing about the manner in which all the main properties of eukaryotic cells could have arisen. The model thus needs at least to be completed. The two theories could be reconciled if the primitive eukaryote envisaged in this chapter happened to derive some advantage from a hydrogen-producing endosymbiotic partner, as is supposed by the new model. Unfortunately, no eukaryotic organism answering this description is known. This does not mean that none ever existed.

Another possibility that deserves to be considered is that the postulated symbiotic association did occur between two kinds of bacteria, as assumed, but took place inside a primitive eukaryote, which somehow benefited from hosting the two partners. Interestingly, an association of this kind actually exists. Some cockroaches harbor in their hindgut a parasitic protist that contains hydrogenosomes and, in close contact with them, endosymbiotic methane-producing bacteria that obviously take advantage of the hydrogen produced by the neighboring organelles. The hydrogenosomes involved in this suggestive threesome have the additional distinction of possessing a vestigial genome.[9]

THE EUKARYOTIC "MIRACLE"

The birth of eukaryotic cells, with all their extraordinary, finely tuned attributes, so different from their "simple"—*tout est relatif*—prokaryotic relatives, is often depicted as the outcome of highly improbable events, one of the major hurdles on the way to humankind, one, perhaps, if the defenders of intelligent design are to be believed, that could not have been overcome without the help of "something else."

This view is understandable; but it is unfounded. Whatever value may be attached to the evolutionary models offered in this chapter, they have at least the merit of showing that the development of eukaryotic cells *can* be explained in terms of natural processes likely to occur when and where they did and to lead to acquisitions retained by natural selection. No doubt, the speculations offered will need to be amended, perhaps abandoned. But their very plausibility should encourage further search for natural explanations. It is far too early to call on "something else."

Also worth recalling is that the single ancestry of the eukaryotic world may not be, as is often claimed, a reflection of its rarity and improbability but could be the simple consequence of a *bottleneck*, a term that has come up several times already in this book. For all we know, the pre-oxygen world may have harbored a wealth of different eukaryotic organisms. Granted that there is no proof of this possibility, it still deserves to be kept in mind. Perhaps some day, molecular sleuthing may become incisive enough to test its reality.

Another point to be remembered from this chapter is that eukaryotic cells are *not* mere associations of bacteria, contrary to an affirmation sometimes made by the unconditional admirers of these organisms (see Chapter 8) and implying, or even stating explicitly, that, since we ourselves are associations of eukaryotic cells, we are "nothing but" associations of associations of bacteria. This view is misleading. No bacteria are known that possess the main features of eukaryotic cells. Any theory claiming to account for the origin of these cells cannot just invoke fateful encounters between prokaryotes. It must provide a plausible explanation for the development of those main features in ancestral cells derived, whether chimerically or otherwise, from prokaryotic cells.

11. The Visible Revolution

THERE IS A GAP OF at least one billion years between the estimated time of appearance of the first eukaryotic cells and the age of the most ancient fossil vestiges of pluricellular organisms. As we saw in the preceding chapter, unicellular eukaryotes, complete with endosymbionts, must have already been present some two billion years ago (and their endosymbiont-less precursors possibly much earlier). Yet, the earliest traces of pluricellular seaweeds are just over one billion years old, whereas the most ancient clearly identified animal fossils so far discovered, known as the Ediacaran fauna, go back little more than 600 million years. There is some evidence that soft-bodied animals, detectable only by the traces they have left in mud, may have existed before that, but probably not earlier than 700 million years ago, at the most.

It would thus seem, barring new evidence to the contrary, that eukaryotic life remained unicellular during many hundreds of millions of years. It did not stay still, however, but fanned out into an extraordinarily rich spread of branches that have given rise to the vast group of organisms formerly classified as protophytes (pro-plants) and protozoa (pro-animals) and now grouped together under the single name of *protists*. Details of this ramification are beginning to be known, thanks to comparative sequencing.

THE EXUBERANT PROTISTS

Open a biology textbook at the chapter, often much too brief, devoted to protists, and you can but marvel at the beauty and efficiency of the forms evolution has carved on the unicellular, eukaryotic model. Protists have not ceased to delight their observers by the multiplicity and wonderfulness of their shapes, specializations, and adaptations. They have names such as amoebae, euglenae, infusoria, diatoms, radiolaria, paramecia, foraminifera, heliozoa, and many others of bewitching poetry. Sometimes visible to the naked eye, many are supported by organic and mineral skeletons of incredible fineness or propelled by motor appendices that delineate exquisite arabesques on their surface.

Many protists, like the diatoms we encountered in Chapter 9, are photosynthetic, distant descendants of the primitive algae that first came to harbor chloroplasts. They are found wherever there is water and light, covering the surfaces of the oceans, seas, lakes, and ponds with a thin sunlight-trapping screen that, directly or indirectly, feeds much of marine life. Others are heterotrophic, admirably organized to catch and digest prey in systems that may reach an incredible degree of complexity, looking like veritable miniaturizations of the digestive tract of animals. These protists truly illustrate the ultimate refinements of the unicellular mode of life. A number of them cause severe diseases such as malaria, sleeping sickness, Chagas disease, and leishmaniasis.

An important group of protists includes organisms of relatively simple design related to fungi, of which *yeasts* are the most famous representatives. Yeasts have been used for several millennia, before their nature was recognized, for such purposes as the making of alcoholic beverages and the leavening of dough, thanks to the two substances they produce by fermentation of sugar: alcohol and, causing the bubbles in the dough and in fizzy beers and wines, carbon dioxide. Yeasts have also played a major role in the modern understanding of life. We saw in Chapter 3 how the discovery that cell-free yeast juice could catalyze the fermentation of sugar struck the death blow to vitalism and, at the same time, launched the unravelling of metabolic processes. The word enzyme recalls these historic events. Yeasts are not only beneficial. Some, like *Candida*, may cause diseases.

Protists, while illustrating the extraordinary complexity single cells can achieve within the framework of eukaryotic organization, also underline the limitations of the unicellular mode, which forces all the functions needed for life to be performed by one and the same cell. The true eukaryotic innovation came when some protists "discovered" the advantages of associating into pluricellular organisms and sharing functional burdens among several different types of specialized cells.

THE MAIN FEATURES OF PLURICELLULAR ORGANIZATION

It is difficult to understand why it took so long for associations to arise that, according to all available evidence, must have been beneficial to their members right from the start. The reason may be that there is something very special, not readily achieved, about associations among eukaryotic cells.

THE SECRET IS DIFFERENTIATION

Association among cells is not a eukaryotic innovation. Many bacterial colonies are known. Some, like the stromatolites (see Chapter 3), are composed of several different species that complete and support each other by their specializations. What was truly new, and perhaps took such a long time to emerge, was the development of such associations from a *single* initial cell, with *all parts having the same genome*. This new creative pathway opened by eukaryotes finally gave life, after some three billion years of hidden existence, the opportunity to inaugurate the most pregnant chapter in its history.

All visible living beings start as a single cell, the fertilized egg. Never mind, for the moment, the fact that this cell itself originates from two cells, the maternal egg cell and the paternal sperm cell. We shall see later the significance of sexual reproduction. The important point is that it all begins with a single genome, which is transmitted in its entirety to the daughter cells at each division. What makes the difference between a simple clone, made of identical cells, and an organism is cellular *differentiation*, that is, the specialization, in different directions, of cells derived from the same parental cell.

With differentiation, a new advantage came to favor cells genetically disposed to group together: *division of labor*. Functions such as external protection, support, motility, digestion, or photosynthesis were carried out by different cells in the association. Thanks to this allocation of tasks, each function became open to improvements that would have been im-

possible within a single cell obliged to accomplish everything simultaneously. Differentiation rules the history of multicellular life. Starting
with two, the number of different cell types has grown progressively, to
reach several tens in plants, and more than 200 in animals. At the same
time, the complexity of the organisms born from this diversification
increased.

But organisms are not just mixtures of cells of different types. Their
cells are arranged into tissues and organs according to a definite blueprint,
characteristic of the species. These dispositions arise in the course of
embryonic development. This is an essential notion for anyone who
wishes to understand the evolutionary history of pluricellular organisms.
Indeed, whenever an organism changes, it is because its *developmental
program* has changed. The mechanisms involved in these phenomena are
beginning to be understood; they imply a true upheaval in the genetic
organization of cells.

SUPERGENES ARE IN COMMAND

We have just seen that the cells of a pluricellular organism all have the
same genes. If they differentiate into distinct cell types, it is because they
do not express the totality of their genes but practice a selection that
varies according to cell type. Otherwise, all the cells of an organism would
be identical. Cells thus contain "genetic switches," systems that switch
on or off the expression of certain genes. This control is carried out by
proteins, called *transcription factors*, that either stimulate or repress the
transcription of the genes involved. These proteins being themselves the
products of genes, which are subject in turn to a similar regulation, genomes house a whole complex and hierarchical network of *regulatory
genes*—the term "supergene" is sometimes used—next to those that code
for "housekeeping," that is, for enzymes, structural proteins, etc.

Regulatory genes are known in bacteria, in which they are involved,
among other things, in the adaptation of metabolism to different nutrients. A historic example, which made the fame of the French investigators François Jacob and Jacques Monod, concerns the manner in which
bacteria transferred to a medium containing milk sugar (lactose) as sole
food supply switch on the genes coding for enzymes specifically needed
to use this sugar. Regulatory genes are, however, much more numerous
in eukaryotes, and their number increases with the complexity of the
body plan of the species. Such is not the case for housekeeping genes,

for which there is hardly any difference among species. Or when there is a difference, impoverishments rather than enrichments most often go together with increasing complexity. Witness the many vitamins we are unable to make, whereas humble bacteria do so without difficulty.

The discovery of regulatory genes has allowed us to discern, at least in principle, the mechanisms that direct and control development. Once fertilized, the egg cell divides into two cells, which similarly divide to produce four, which divide into eight, and so forth. Soon, in the course of this process, the cells cease to be identical. Depending on their location in the ensemble, they start expressing or stifling certain regulatory genes, with the consequence that the proteins translated from those genes create concentration gradients between the areas where they are produced and those where they are not. These gradients influence in unequal fashion the expression of other genes, which, in turn, influence others, in a cascade whose complexity soon exceeds the limits of our imagination and even anything that can be simulated by the most powerful computer programs. At the end of the game, there is an oak plantule, a jellyfish larva, or a newborn baby, depending on the program written into the genome.

Such a mechanism has long been suspected. Already, in the early part of the twentieth century, the German embryologist Hans Spemann demonstrated, by means of remarkably skillful and ingenious experiments, the existence of what he called morphogenetic—shape-creating—gradients in embryos. Modern biology is beginning to flesh out those gradients in terms of genes and their protein products. Particularly important has been the discovery of so-called *homeogenes*, whose control is so wide-ranging that a single mutation of such a gene may cause a fruitfly to grow an extra pair of wings or to sprout additional antennae on its head. Homeogenes have been recognized throughout the pluricellular world, from simple fungi to the most complex of animals.

EVOLUTION OCCURS BY WAY OF DEVELOPMENTAL PROGRAMS

With these elementary notions, we may now address the problem of evolution, which is conditioned, as we have seen, by changes in the developmental program of organisms. This fact implies almost necessarily that the underlying genetic changes have as targets regulatory genes. But all depends on the cell type to which the modified gene belongs. Thus, a mutation in a skin, stomach, or brain cell may start a new cell line, for example, a cancerous one. But the individuals concerned do not, if they reproduce, give birth to descendants afflicted with cancer of the skin, stomach, or brain. Only genetic modifications of a germ cell that will eventually be involved in the generation of a new individual can be of significance for evolution. Such modifications are the only ones that can influence the development of the fertilized egg. They are also the only ones that can be hereditarily transmitted, as they affect all the cells of the organism, including those that will become germ cells in turn and will give rise to the next generation.

A second consequence of the new evolutionary mode is that a delay, increasing with the complexity of the developmental program, may separate a genetic modification from its effect. In the case of bacteria, a mutation has an immediate effect on the ability of the affected cell to survive and multiply. But, for a pluricellular organism, the effects of a genetic modification on the key criterion, which is the ability to produce progeny, can be evaluated by natural selection only after sexual maturity has been reached, at least if the effects are to be beneficial. If they are harmful, they can, of course, become manifest at any stage. The relationship between cause and effect is rarely simple, as a given genetic modification may affect different cell types in very different ways. This has become increasingly evident with the creation of transgenic organisms, that is, organisms stemming from a genetically modified egg. In spite of the advances of biotechnologies, we are still far from mastering evolution.

A third factor of primordial importance for our understanding of the evolution of pluricellular organisms is that they reproduce *sexually*. Of what does this reproductive mode consist? What are the advantages that have caused sexual reproduction to be almost universally retained, in animals as well as in plants and fungi? How does this process affect evolutionary phenomena? These questions will be addressed in the following pages.

SEX IS A POWERFUL AUXILIARY OF EVOLUTION

We all have two parents. With rare exceptions, this is true, at least at the cellular level, of all pluricellular organisms. All, or almost all, arise from the fusion between two cells. Even unicellular organisms sometimes join to reproduce. The main consequence of this system, and probably also its greatest advantage, is that we receive one half of our genes from our mother and the other half from our father. We are like our parents but identical to neither. Sexual reproduction is a source of *genetic diversification* within the same species, allowing it to adjust more readily to environmental changes.

But this is not all. Sexual reproduction implies that each individual has *two complete genomes*. Indeed, the germ cells that conjugate do not each possess half a genome. Germ cells are perfectly viable and may even give rise to complex organisms (in plants, for example); they have a complete genome, borne by a single set of chromosomes, and are termed *haploid* (from the Greek *haplous*, simple) for this reason. The cells born from the conjugation of germ cells have two complete genomes; they are called *diploid* (from the Greek *diplous*, double). Behind sexual reproduction, there is diploidy, with, as a concomitant, *gene duplication*.

This is an ancient phenomenon. There is evidence of gene duplication having occurred even before the emergence of the last common ancestor of the living world. The advantage of the process is obvious. One of the two genes of the pair may undergo all sorts of mutations while the other goes on doing its normal job. Evolution can thus "experiment," without risk, with diverse variants of the gene, until a variant appears that proves useful and becomes a new gene. The study of sequences shows that a great many genes have arisen in this manner.

Genetic diseases illustrate clearly the benefits of diploidy. We all own several tens of abnormal genes that would have brought about our early demise were they not accompanied by a healthy homologue. If, by chance, both parents contribute the same abnormal gene to the fertilized egg, the genetic defect becomes manifest.[1] Such coincidences are fortunately rare, but their risk increases with the degree of genetic similarity between the two parents, as is shown by the nefarious consequences of inbreeding.

We have seen that germ cells have a single set of chromosomes. The formation of germ cells from cells that have inherited two sets of chromosomes from the fertilized egg thus implies the loss of one of the two sets. This *reduction of the chromosome number* takes place in the course of germ cell maturation, in a special kind of mitotic division called *meiosis*, in which the cells divide twice while the chromosomes are duplicated only once. The germ cells thereby each emerge equipped with a single set of chromosomes, which, however, is not a simple heritage of one of the two sets, of maternal or paternal origin, that were present in the parental cells.

Indeed, in the course of a phase of meiosis, the chromosomes of each pair align intimately against each other and exchange segments of homologous DNA by what is known as *crossing-over*. As a result of these exchanges, which are called genetic *recombinations*, each chromosome becomes a mosaic of genes of maternal and paternal origin. The homologous chromosomes reorganized in this manner are then randomly assorted into the two complete sets that are assigned to the two daughter cells. It follows that each germ cell possesses a different version of the characteristic genome of the species. Each individual stemming from the union of two germ cells is *guaranteed to be unique*. This is the source of our individuality. Only true twins, originating from the same fertilized egg cell, are genetically identical. This is why they often serve in studies aimed at identifying the respective parts played by heredity and environment in the determination of certain traits.

In most organisms, germ cells, or *gametes* (from the Greek *gamos*, marriage), have very different shapes in the two sexes. The female gamete, or egg cell, is an immobile cell of large size, frequently loaded with reserve substances. The male gamete, or sperm cell, is small and most often consists of little more than a membrane-enclosed nucleus containing the genetic material and equipped in animals and lower plants with a propelling flagellum. Fertilization occurs when the sperm cell nucleus penetrates the egg cell. As a result of this mechanism, the fertilized egg cell's cytoplasm is provided almost exclusively by the female gamete. Mitochondria, in particular, are thus transmitted by the female line. We shall see that this fact has allowed an interesting method of phylogenetic reconstruction based on the sequencing of mitochondrial DNA (see Chapter 16).

The organs in which gametes are formed may belong to the same individual, then called hermaphrodite, or to different individuals. In the latter case, typical of all vertebrates, differences, which may be spectacular, usually distinguish males from females. Called *sexual dimorphism*, this distinction is linked to a special pair of chromosomes, called sex chromosomes, in a manner that varies with species. In the human species, as in other mammals, the female individual possesses two homologous chromosomes, of formula XX. In the male, one of the X chromosomes is replaced by a Y chromosome. In the course of chromosome reduction at meiosis, half the sperm cells inherit an X chromosome, and the other half a Y chromosome. The sex of the offspring is decided at fertilization. Depending on whether the fertilizing sperm cell possesses an X or a Y chromosome, the fertilized egg will generate a female (XX) or a male (XY) individual.

Men differ from women in that they possess only a single copy of a number of genes borne by the sex chromosomes. They do not, for these genes, enjoy the benefits of duplication. This drawback of the male condition is illustrated by the genetic deficiencies localized on the X chromosome, of which hemophilia, a blood coagulation defect, is a classic example. Transmitted by the mother, who is most often unaffected because she is protected by her second X chromosome, these diseases strike boys selectively, with one chance in two, depending on whether the X chromosome supplied by the mother to the fertilized egg is, or is not, the bearer of the deficiency.[2] This discrimination against males is not a law of nature. In many animals, for example, birds, the female has the disadvantage of two different sex chromosomes. In addition, evolution has, for obscure reasons, deprived mammalian females of part of their advantages. In differentiated cells, one of the two X chromosomes, apparently selected at random, becomes inactive.

In conclusion, the main advantage of sexual reproduction is a continual reshuffling of genes, which diversifies individuals and gives species greater flexibility in adapting to changing environmental conditions. Moreover, thanks to the fact that genes are present in two copies, sexual reproduction allows giving a try to all kinds of genetic variants, possible sources of new evolutionary forms, without compromising the viability of the species. But here there is a complication. In unicellular organisms, each mutation creates a new lineage. In pluricellular organisms, mutations merely introduce new genes into a common melting pot, called the *gene pool*, that is being continually churned by sexual reproduction. The manner in which new species may emerge is still being debated. The general belief is that speciation requires the variant form to remain geographically isolated long enough to evolve to a stage where hybridization with the unchanged form has become impossible or, at least, produces an infertile form, such as the mule, the sterile product of the mounting of a mare by a donkey. Apparent exceptions to the condition of geographical isolation have, however, been observed.[3]

A Bird's-eye View of Pluricellular Evolution

Retracing in detail the evolution of pluricellular organisms does not belong in this book. Any elementary biology textbook summarizes what is known of this history. I shall provide only a very brief overview. As will be shown in the next section, the evolution of reproductive strategies provides a highly revealing guideline to the underlying selective mechanisms.

PLANTS

According to present knowledge, it all started about a billion years ago, perhaps earlier, with an association of some *unicellular green algae* giving rise to the first *seaweeds*, which diversified into the many species found in the aquatic vegetations of today. From some ancestral seaweed there then emerged primitive *mosses*, equipped with minimal properties allowing them to survive on land: a water-impermeable skin, or cuticle, protecting against desiccation, with small holes, or stomata, allowing exchanges of carbon dioxide and oxygen with the atmosphere; and rootlike structures capable of drawing from the soil the necessary water and mineral elements. The next step in the invasion of land by plant life saw the constitution of vertical polarization between roots and leaves, with the development, between the two, of a dual vascular system conducting toward the leaves the mineral sap drawn from the soil by the roots, and toward the roots and the other nonphotosynthetic parts of the plant the organic sap formed in the leaves with the help of sunlight. With the appearance of lignin, a particularly resistant substance, the first arborescent plants were born, eventually leading to the luxuriant marshy forests of the Carboniferous era (360–286 million years ago), whose fossil remains now fuel a good part of our energy needs. Ancestral, among others, to present-day *ferns*, many of these organisms fell victim to cold and drought during the great Permian crisis (286–250 million years ago), which decimated a large number of living species. They gave place to gymnosperms, of which *conifers* are the main extant representatives, which resist climatic rigors better. Then came the angiosperms, or *flowering plants*, which now dominate the scene.

ANIMALS

The first animals arose in water from heterotrophic protists as simple associations, which, starting with only two kinds of diffentiated cells, progressively developed a number of distinct organs allowing the animals to move, eat, digest, respire (oxygen dissolved in water), excrete, and reproduce, all coordinated by a growing network of nerve cells linking sensory and motor areas and joined by an internal milieu that eventually came to be distributed throughout the organism by a circulatory system. This assemblage was completed by the appearance of humorally transmitted regulatory substances (hormones) and by the beginnings of an immune system. This is the basic body plan of animals.

Intermediates in this long history have left descendants in the sponges, jellyfish, sea anemones, corals, polyps, worms, and other members of the rich and colorful world of *lower marine invertebrates*. Then, a capital event occurred: *segmentation*, the result of duplication of particularly important regulatory genes. The organism became a chain of semi-independent organisms, each almost identical to the other segments at the start, as is the case, for instance, in the common earthworm. We have already seen the evolutionary importance of duplication, first of genes and then of entire genomes. With body duplication, a veritable riot of evolutionary experimentation was inaugurated. Next to segmented worms there arose the huge division of *arthropods*, including crustacea, insects, arachnids, and many other groups, in which the evolution of each segment in a different direction is readily recognized. In the likewise important group of *molluscs*, segmentation has suffered so many modifications as to be hardly discernable, except in primitive forms, such as tritons.

While this diversification was taking place, another sudden change created a major split in the history of animals. Whereas, so far, the mouth end of the alimentary canal had been derived from the blastopore, the single opening of the early embryo, and the anus had appeared later, this situation was reversed in a mutant form, generating the new group of *deuterostomes* (mouth second), as opposed to the existing *protostomes* (mouth first). This line gave rise to the sea urchins, starfishes, and other members of the *echinoderm* family, which have evolved into circular five-segmented forms. Especially, it has produced the *chordates* and, through these, the *vertebrates*, in which segmentation has constructed, first with cartilage and later with bone, the articulated framework of the spine. *Fish* were born.

In the meantime, plants had spread inland, offering rich pastures to the animals that succeeded in surviving and reproducing. To achieve this, the animals had to protect themselves against desiccation, develop adequate means of locomotion, and acquire the ability to utilize atmospheric oxygen. Worms, molluscs, arthropods, and fish each "found" different solutions to these problems. In fish, in particular, fins became legs and primitive lungs, believed by some to be ancestral to the swim bladder, turned into indispensable respiratory organs. *Amphibians* represent the first fruits of these adaptations. From this group arose the *reptiles*, made famous by the dinosaurs. Further evolution of the reptiles produced two main branches, one leading to *birds* and the other to *mammals*. The latter fanned out into a large number of species, among which the *primates* developed, a bare six million years ago, the branch that led to the *human* species.

FUNGI

Fungi, or mycetes, form a third major group of pluricellular organisms, which were long believed to be related to plants because, like plants, they are immobile and encased within rigid external structures. Molecular sequencing, however, has shown that fungi are phylogenetically closer to animals, with which they share a heterotrophic mode of life, than they are to the autotrophic plants. Some fungi pursue a hidden existence in the depths of the soil, where they weave invisible filamentous networks that may cover several square kilometers. Then, suddenly, there emerge from these networks remarkable structures of varied shapes and colors, *mushrooms*, whose main function is reproductive.

A remarkable feature of the complex history of pluricellular life is the important role, possibly decisive, played by improvements in reproductive mechanisms. This point is important enough to deserve some elaboration.

REPRODUCTIVE SUCCESS IS A DRIVING FORCE OF EVOLUTION

The evolutionary success of a species depends on the ability of its members to reproduce. The fact is evident, but it has needed Darwin to underline its importance and significance. We are not dealing here with a mere logical necessity, sometimes depicted as tautological by the adversaries of Darwinism; it is a fact, evidenced by a remarkable trend in evolution towards increasingly efficient reproductive mechanisms. This pathway can be reconstructed, for it has left traces. Species of more ancient origin continue to occupy niches where their reproductive means remain adequate in spite of their primitive character. What reproductive improvements have allowed is the conquest of new environments or, alternatively, the ability to withstand climatic and other rigors that more poorly equipped species could not survive.

The success of sexual reproduction is contingent on the favorable realization of three phenomena. At the center, there is *fertilization*, which depends on the successful encounter of male and female gametes. This act is preceded by *maturation of the gametes*, with, as a key step, chromosome reduction; it is followed by the *development* of the fertilized egg up to the stage of autonomous survival and reproductive potential. Plants and animals follow this scheme differently.

PLANT REPRODUCTION

Contrary to animals, in which only cells of the germ line have a single set of chromosomes (haploidy), all other cells in the body having two sets of chromosomes (diploidy), plants often go through an important haploid stage, which may extend to the point of dominating development. Chromosome reduction leads to a first kind of germ cells, called *spores*, which give rise to a more or less developed form of the plant, composed of cells that possess only a single set of chromosomes. From this form are born the *gametes*, which conjugate to make the fertilized egg. A second form, with cells possessing two chromosome sets, then develops from the latter and, after chromosome reduction, produces spores. It then starts all over again.

This reproductive mode is called *alternate generation*. It brings about the alternative succession, by way, in turn, of spores and of conjugating gametes, of two forms, one endowed with a single set of chromosomes and another with two sets. Evolution has composed many variations on the basic theme of alternate generation. Among seaweeds already, there are, next to species in which the two generations have the same importance and almost the same appearance, many variants in which one or the other is dominant.

For plants, which cannot move, spores are the main vehicle of dissemination. Spores have played an essential role in the adaptation of plants to land life, thanks to their acquisition of a resistant and impermeable shell allowing them to be dispersed by air and to await, in the soil, conditions propitious for germination. This kind of phenomenon is not a eukaryotic innovation. Many bacteria, when exposed to unfavorable conditions, transform into resistant spores capable of subsisting for a very long time. The "cursed fields of Beauce," sites of a celebrated vaccination experiment against anthrax by Pasteur, owed their curse to the presence of bacterial spores that remained infectious for cattle even after several years.[4] The ability of bacteria to sporulate is likewise responsible for the fact that soil-stained wounds expose us to the risk of tetanus.

Helped by their spores, primitive plants have succeeded in progressively migrating out of their original watery milieu and reproducing on land. Mosses, for example, produce spores able to await, under conditions of dryness, the humidity that will allow germination. Subsequent development results in a complete organism whose cells contain a single chromosome set and from which, eventually, the gametes arise. After fertilization, the egg cell produces a transient stage, characterized by two chromosome sets. In this stage, which is of short duration, meiosis soon takes place and spores are produced and shed.

The opposite situation exists in ferns, in which the dominant form has two chromosome sets. Spores go through a discreet subterranean developmental phase, which leads to gametes that rapidly fuse. The egg thus fertilized gives rise to the visible plant, which, at the end of the season, produces the spores that will ensure the new generation.

At first, there was only a single variety of spores and, thus, a single form producing the two types of gametes. These, being born at the same site, had no difficulty meeting. An important new step was accomplished with the differentiation of spores into male and female types, from which sperm cells and egg cells were to emerge separately. This distancing made fertilization more uncertain but had the advantage of favoring the encounter between genetically different gametes.

In a subsequent stage, female spores remained on the plant, to pursue their maturation within an organ, the *ovary*, which surrounds the resulting egg cells with a protective and nutritive envelope. Henceforth, only male spores served in dissemination; they are contained in the *pollen* grains. In order to germinate and turn into male gametes, these grains must enter a compatible ovary, which alone provides adequate conditions. The enormous number of pollen grains produced compensates for the chanciness of this mechanism.

In such plants, fertilization takes place inside the ovary and is followed by a partial development of the fertilized egg. The process then is arrested. The ovary closes around the early embryo and becomes a *seed*, which detaches from the plant and remains in the soil in a dormant state until local conditions allow the embryo to resume its development. Plants that reproduce in this way are called *gymnosperms*, meaning with naked seeds (*gymnos* means naked in Greek). Conifers are their principal representatives. This new reproductive mode proved its superiority during the Permian crisis, when, as we have seen, the great Carboniferous forests were decimated.

In a last development, the ovary and the pollen-producing organs transformed into structures of extremely varied shapes, colors, and scents, the *flowers*. These attracted insects and other animals, which thus came to play a dominant role in pollination. An additional innovation led to the construction around the seeds of complex envelopes richly supplied with nutritive substances. They are the *fruits*, whose diversity rivals that of the flowers. Plants whose seeds are embedded within fruits are called *angiosperms* (from the Greek *aggeion*, envelope). They represent the majority of plant species in the present-day world.

Thus, from spores to seeds and from seeds to flowers and fruits, plant evolution has given rise to increasingly perfected reproductive machineries that have allowed primitive seaweeds to invade the lands and to adapt to a wide variety of climatic conditions by a pathway that has left, as main witnesses of its different steps, the mosses, the ferns, the conifers, and, finally, the whole exuberant gamut of flowering plants.

ANIMAL REPRODUCTION

Animal evolution illustrates a similar trend toward increasingly complex reproductive mechanisms. As long as animals stayed confined in water, their reproduction remained largely aquatic. The females of marine invertebrates and even of fish lay unfertilized eggs in water, and males discharge their sperm cells in the vicinity. Fertilization takes place in water and the fertilized egg develops in the same milieu. At first, when amphibians started to move out of the sea, the ancestral mode of aquatic development continued to be followed. Frogs, for example, depend on external fertilization (even though they may mate), and their fertilized eggs develop in water until metamorphosis of the tadpoles. A few cases are known in which the developing embryos are sheltered in cavities of the female or, sometimes, the male body.

In a number of cases, in both vertebrates and invertebrates, the ejected sperm cells find their way into the female body and fertilize an egg cell locally, resulting in a fertilized egg that starts developing internally. This happens, for example, in viviparous fish. Internal fertilization became the rule with the advent of *copulation*, which, by bringing sperm cells into the immediate vicinity of egg cells, greatly decreased the chancy character of aquatic fertilization and, at the same time, provided a favorable milieu for early embryological development. This was an essential factor in the further evolution of animals and in their adaptation to land life.

A major step of crucial importance was the appearance of the *amniotic egg*, which signalled the complete liberation of vertebrates from their original aqueous milieu. Water remains essential for development of the embryo, but it is no longer the water of a pond, sea, or ocean. It is the amniotic fluid, contained within a pouch surrounded by a protective shell, where the embryo can develop using the nutritive reserves stored in the egg. This is the characteristic reproductive mode of reptiles, birds, and the first mammals, the monotremes, of which the platypus is an extant representative. Monotremes continue to lay eggs, like their ancestors.

Finally came *intrauterine* development of the embryo, a mode occasionally observed already in lower animals but highly perfected in the mammals. In marsupials, such as koala bears and kangaroos, the intrauterine phase is still short and largely dependent on the nutritive stores of the egg, to be followed subsequently by a long period of milk-fed maturation within the *marsupial pouch*. In the placentals, which comprise the bulk of modern mammals, the ultimate improvement was brought about by the development of the *placenta*, which allows the embryo to be nourished directly by the mother's blood.

The evolution of land arthropods, notably insects, illustrates a somewhat similar history. Development is aquatic in a number of species, for instance, mosquitoes. It is terrestrial in many others, often linked with instinctive behaviors of high complexity that ensure the protection of the eggs during embryonic development.

Thus, the evolution of animals, like that of plants, is landmarked by improvements in reproductive strategy. Vertebrates, in particular, starting from eggs fertilized and developing in water, went over to fertilization by copulation and to amniotic eggs capable of independent terrestrial development, then to intrauterine eggs with largely marsupial development, and, finally, to placental eggs developing entirely within the womb.

REPRODUCTION IN FUNGI

In fungi, reproductive mechanisms are more diverse than complex. Some organisms have even retained the ancestral, asexual mode of reproduction. The others multiply by a variety of sexual mechanisms that often, as with plants, depend on *spores* for dissemination. Many molds grow on the surface of their nutritive support and leave it to the wind to disperse their spores. It was such a spore that, by contaminating a microbial culture in the London laboratory of the Scottish bacteriologist Alexander Fleming, led to one of the greatest miracle drugs in the whole history of medicine.[5] The main reproductive innovation in fungi is represented by *mushrooms*, which emerge from the soil to disperse the spores of many underground organisms.

COMMON STRATEGIES

It is remarkable that, in spite of very different modes of life, plants and animals have developed *closely similar reproductive strategies*. In both, male forms soon found their role reduced to providing a second genome to the egg cell, with the help of either actively motile or passively transportable "disposable" sperm cells. In both, evolutionary innovations have mostly concerned the female forms, supplying them with increasingly elaborate mechanisms for attracting the sperm cells, favoring fertilization, and sheltering the developing embryo. Seeds and fruits, on one hand, amniotic eggs, first external and later internal, on the other, are the most sophisticated outcomes of this trend. The most important difference between the two groups lies in the utilization, by plants, of spores and of more or less developed haploid forms. These intermediates do not exist with animals, which rely on their own mobility to ensure their dissemination.

The Development of the Nervous System Traces the Course of Animal Evolution

There is a second privileged evolutionary direction: that leading to nervous systems of increasing complexity in the animal line. With the exception of a few primitive animal forms, such as sponges, all animals possess nerve cells, or *neurons*. These cells are characterized by two kinds of extensions, acting as receivers (dendrites) and emitters (axons) of signals. They can thus serve to connect *sensory* organs reacting, for example, to light, sound, or touch, with *motor* organs capable of initiating an appropriate movement, such as the capture of prey or flight from a predator. A particularly important property of neurons is that they can *join together* into signalling chains. Especially, they can associate laterally with such chains in a manner that allows modulation of the transmitted signals. The development of increasingly complex polyneuronal networks is one of the main lines of animal evolution.

A jellyfish has a simple necklace of neurons, which is enough to cause its muscles to contract in coordinate fashion and propel its body. Of the 959 cells that compose a particularly well-studied, minuscule worm of the class of nematodes, *Caenorhabditis elegans*, 302 are neurons, illustrating the importance of these cells. With a brain no bigger than a pin's head, bees build the honeycombs of their hives and leaf-cutting ants cultivate the fungi on which they feed. What we derisively qualify as a bird's brain allows a pigeon to find its way back home, without seeing the sun, over a distance of more than 1,000 kilometers, and the bird of paradise to court the mate of its choice by a complicated choreography on a scene it has previously decorated with branches and flowers.

With mammals, brain and behavior became even more complex, reaching a particularly remarkable degree in certain primates, such as chimpanzees, and also in some cetaceans, such as dolphins. Then, in a primate line closely related to present-day chimpanzees, the brain suddenly began developing at an extraordinary pace, to the point of reaching, in less than six million years, *three times* the size it took 100 times longer to acquire before that. This astonishingly fast development went together with the manufacture of increasingly sophisticated tools, shelters, and machineries. It opened into the paintings of Lascaux, the Iliad, the Bible, the well-tempered clavier, the theories of relativity and natural selection, nuclear energy, genetic engineering, and artificial intelligence. At the same time were born the sense of good and bad, the search for truth, the emotion before beauty, the feeling of love, metaphysical anxiety, and apprehension of the mysteries of nature and its origins to which humans gave the name of God. I shall come back to this at length in subsequent chapters.

Thus, from the necklace of half a dozen neurons that circled the orifice of a primitive jellyfish to the collection of some 100 billion neurons that make up the human brain, the nervous system has, over 600 million years, steadily progressed towards increasing complexity. All along this road, side paths have detached, stopping at less-developed brains sufficient for the needs of their owners and allowing these to diversify into numerous secondary ramifications. The privileged direction, nevertheless, remains evident.

12. The Arrow of Evolution

As related in the preceding chapters, we finally have an answer to a question—where do we come from?—that has haunted humans ever since they gained the capacity to wonder. We now know that our origin is lost in the darkness of time. It goes back to such a remote past that our minds fail even to imagine it. Almost four billion years, four million millennia, 40 million centuries! Just to count them, at the rate of one century per second, we would need more than a year. In those immensely distant times, there appeared on our young planet, freshly recovered from the violent impacts and convulsions that heralded its birth, the primitive form of life from which, with all other living beings on Earth, we have descended. The line that links us to this ancestral form is uninterrupted. But it can be recognized only retrospectively.

The Meandering Road to Humankind

During more than two billion years, life wove its occult networks of unicellular organisms over the surfaces of lands and seas, in the dark depths of oceans, and in the hidden crevices of rocks, radically upsetting the natural balances that governed the planet and recycling the elements through a new chemistry, still subject to the laws of nature, but channelled along strange pathways by molecules of its own making that turned into its rulers. From this pullulation, there emerged, through an astonishing metamorphosis, the eukaryotic cells that allowed life to escape from its prokaryotic shackles.

Surprisingly, it took these cells more than one billion years to discover the new, collaborative mode of existence that was to prove so enormously successful. It is only after life had already covered the major part of the distance that separates its beginning from the present time that eukaryotic cells born from the same parental cell started to form organisms in which vital functions were distributed among different cells. Next to plants and fungi, pluricellular, aerobic, heterotrophic, mobile living beings arose: the animals.

In the thick bush of evolutionary ramifications that grew from the first animals, a line can be distinguished *a posteriori*. It first goes through a string of sponges, jellyfish, polyps, and marine worms. Then, letting those primitive forms fan out laterally with all the invertebrates that arose from them, including crustacea, insects, molluscs, echinoderms, and many others, the line leads to the first vertebrates. After that, leaving the fish in the seas, it comes out of the water with the first amphibians, follows some of these in their transformation into reptiles, and pursues its progression up to the early mammals, letting amphibians and reptiles, together with the birds that emerged from the latter, diversify on their own. Among the mammals, the line comes to the primates and continues its course, by way of the big anthropoid apes, up to a species close to present-day chimpanzees. From this stage, reached some six million years ago, a final stretch leads, by way of a series of intermediary forms paleoanthropologists are beginning to identify, to the modern human species: *Homo sapiens sapiens*.

This long, tortuous evolutionary pathway leading from the first ancestral form of life to the human species is landmarked by a large number of *bifurcations*, or *forks*, each of which signals an evolutionary step where our line diverged from another that either died out or branched into some other living group. How many such bifurcations there were is difficult to estimate, but they must have numbered at least in thousands. As we have seen in Chapter 7, modern science tells us that each of those bifurcations was the consequence of an accidental genetic modification that happened, by chance, to occur in an environment favorable to the selection of the mutant form. According to this description, we are the outcome of *thousands of fortuitous events*, each of which has involved a dual role of chance, at the genetic and at the environmental level. This much is accepted by a vast majority of scientists. The significance of this fact is, however, diversely appreciated.

Who Set the Itinerary?

For a number of experts, perhaps the majority, the message from science is inescapable. It was all an incredible piece of luck. The emergence of the human species was an extremely improbable event, so improbable as almost surely to be unique. Even if there should be life in other sites in the universe, which is far from certain, the likelihood of its leading to human beings or even to any conscious, intelligent, humanlike beings, is vanishingly small. The event, we are told, is so unlikely that it might very well not have happened on our planet either, where only an extraordinary combination of circumstances made it possible. This view, which has been expressed in various terms by such leading evolutionists as George Gaylord Simpson,[1] Ernst Mayr,[2] Stephen Jay Gould,[3] and many others, was summed up by Monod, when he completed the sentence cited in Chapter 3, "the Universe was not pregnant with life," with "nor the biosphere with man."[4]

A small but increasingly vocal minority goes one step further. Initiated in France, in the wake of Teilhard de Chardin's evolutionary mysticism, this movement has also gained the Anglo-Saxon world. The emergence of humankind, according to these modern defenders of finalism, is not just highly improbable, it is plain impossible, at least by way of strictly natural processes. There are just too many unlikely coincidences piling up. Without outside help, chance "wouldn't have a chance," however many opportunities the universe, or even trillions of universes, would supply. Sum up the thousands of fortuitous occurrences—which means multiplying their probabilities—that traced the road from the ancestral form of life to humankind, and you have long passed the point where improbability turns into impossibility. Yet, it happened. So, there must be "something else."

This cryptic entity has been ascribed to some yet-to-be-discovered principle of unknown nature, vaguely described by words such as autopoiesis, self-organization, complexity law, or informational force, or even identified as a mysterious manifestation of quantum mechanics. Some do not hesitate to see it as the hand of God actually manipulating genes.

We have already seen, in Chapter 3, how intelligent design has been invoked by Behe to explain the "irreducible complexity" of biochemical systems. The same kind of argument, based on "integrative complexity," is applied to biological evolution by the New Zealand scientist Michael Denton in his 1998 book *Nature's Destiny*, significantly subtitled "*How the Laws of Biology Reveal Purpose in the Universe.*"[5] Denton accepts the Darwinian explanation and does not explicitly call on "something else," but he comes perilously close to it, with words such as "directed evolution,"[6] "ingeniously contrived,"[7] or "preordained."[8] In discussing the development of the characteristic lung of birds, he finds it "hard not to be inclined to see an element of foresight in the evolution of the avian lung, which may well have developed in primitive birds before its full utility could be exploited."[9] The whole blueprint, he believes, is written in detail into the DNA, presumably by a Creator who knew what He was doing and where He was going.

Divine rigging or intervention in evolution is not, however, seen by all believers as a mandatory correlate of religious faith. In *Finding Darwin's God*,[10] a book written largely in response to Behe's *Darwin's Black Box*, the American biologist Kenneth Miller, who, like Behe, identifies himself as a practicing Catholic, gives a detailed and incisive critique of Behe's arguments and cogently exposes their fallacy. Miller himself has no problem accepting the neo-Darwinian theory in all its aspects, including the utter contingency and unpredictability of the human species. To Miller, in fact, contingency is part of God's plan. It is written into the laws of physics and forms a necessary condition for human beings to be free to accept or refuse God's will.

It should be added that not all scientists agree with the basic premise that the origin and evolution of life were highly improbable events that, to take place, required a nudge either from chance or from a Creator. Some, mostly astronomers or cosmologists, go so far as to claim that the probability of intelligent life arising naturally is so great that a serious effort deserves to be made to establish contact with at least one of the many civilizations that must exist "out there." As we shall see in Chapter 17, holders of this view have made their case with sufficiently convincing vigor to gain the commitment of considerable resources to a search for extraterrestrial intelligence (SETI), which aims at detecting signals sent by some distant civilization.

The issue, clearly, hinges on a *quantitative* estimate of probability. Depending on the value you assign to the likelihood of the overall process that has led to humankind, you are in the camp of the "impossibilists," the "improbabilists," the "probabilists," or the "obligatorists." Simple gut feelings are not enough for such a decision. We must look more closely at the notion of chance itself, as it applies to evolutionary phenomena. Chance does not operate in a vacuum; it is always subject to constraints.

The Constraints of Chance

Chance does not mean an unlimited number of possibilities. It simply signifies that the choice among existing possibilities is *governed strictly by their probabilities*. The number of possibilities is always finite. At heads or tails, it is two. At dice, it is six with one die and 21 with two dice.[11] At roulette, it is between 36 and 38 depending on the number of zeros. At a lottery, it is ten million if the numbers have seven digits. At the game of bridge, it is 5×10^{28} (50 billion billion billions). The number of possibilities may be very high, but it is never infinite. This means that a given result can always be achieved with near certainty, even though it depends strictly on chance, if chance is solicited often enough. This, of course, has to be physically possible. Within realistically acceptable limits, all depends on the number of opportunities chance is offered to produce a given event, as compared to the probability of the event.

Take a perfect coin and let it be tossed by a robot strictly ruled by the laws of chance. The probability of the coin falling on the heads side is one in two. Have it tossed a second time, and the probability of its falling on the heads side is again one in two. However, the probability of the coin's falling on the heads side twice in succession is $1/2 \times 1/2$, that is, one in four. It follows that, with two tosses, the coin has three chances out of four of falling at least once on the tails side. With three tosses, the probability of its falling each time on the heads side is $1/2 \times 1/2 \times 1/2$, or one out of eight. Hence, the coin has seven chances out of eight of falling at least once on the tails side. Pursuing this kind of reasoning, one readily finds that it suffices to toss the coin ten times for the probability of its falling at least once on each side to be 99.9 percent. We approach certainty.

One easily calculates in the same manner[12] that a 99.9 percent probability of winning a bet is reached by throwing a die 38 times or by giving a roulette wheel about 250 spins. Even a seven-digit lottery number has a 99.9 percent chance of coming out if some 69 million drawings are done. The number of times the cards must be dealt at bridge to guarantee a given distribution with the same probability is likewise finite and theoretically computable, even though it is beyond the capacities of my hand calculator and, also, beyond the boundaries of any realistic physical possibility.

In conclusion, *chance does not exclude inevitability*. The probability of an event, however improbable, can become close to certainty if the event is given a sufficient number of opportunities to take place. This is of little help to those who play games of chance, because the games are always rigged so as not to give chance enough opportunities. But it may be relevant to the evolutionary game.

MUTATIONS ARE RARELY THE LIMITING FACTOR IN EVOLUTION

The number of mutations a given genome can undergo is large, but it is finite and, in many cases, not excessively large relative to the number of individuals at risk—often millions, if not billions or more—and to the time—up to millions of years—evolution has available for its experiments. Contrary to what is often assumed, evolution rarely has to wait very long for some favorable mutation to occur. More often than not, mutations are present in a population all the time or come up regularly, waiting, so to speak, for the environment to provide them with an opportunity to prove useful.

This assertion is easily demonstrated in the case of simple point mutations, that is, those in which one base is replaced by another. Considering only replication errors, which are known to occur with a frequency of about one wrongly inserted base in one billion, it is readily calculated[13] that the probability of finding a given point mutation in a clone of cells produced by successive division from a single cell becomes 99.9 percent after about 34 generations, that is, less than one day for bacteria, and about one month for animal cells.[14] To take a more concrete example, consider red-blood-cell renewal in the bone marrow of an adult human individual. The probability of a given point mutation taking place in the course of this process reaches 99.9 percent after only about two hours. It is fortunate that most of those mutations are harmless and that, in addition, evolution has provided elaborate repair mechanisms to offset their effects. There are, however, rare cases in which an accident of this sort may lead to some anomaly, such as cancer.

It may be argued that such calculations are of limited value because point mutations are likely to be of little importance for evolution, which probably depends mostly on more extensive genetic rearrangements whose probability is much more difficult to evaluate. It is therefore significant that all we know of evolution in action tends to confirm the richness of the mutational field.

Take, for example, the many cases of *drug resistance*. In only a few decades, the appearance of "superbugs" resistant to penicillin and most other antibiotics has grown into a worldwide health problem. One could argue that bacteria, with their huge numbers and fast multiplication rates, make a special case. But the problem is not limited to prokaryotes. The malarial parasite, *Plasmodium falciparum*, which is a eukaryotic protist, has become largely resistant to chloroquine, the drug that was produced during the last war as a substitute for the no-longer-available quinine, and it is now rapidly developing resistance against mefloquine, the latest antimalarial created in the 1960s by the U.S. Walter Reed Army Research Institute. Even more impressive, in 1948, the Nobel prize in *medicine* was awarded to the Swiss chemist Paul Müller "for his discovery of the high efficiency of DDT as a contact poison against several arthropods." Malaria, it was triumphantly announced, was going to be vanquished by the simple means of killing off the mosquitoes that transmit the parasite.[15] Today, the use of DDT is increasingly being questioned, not only because of its noxious effects on the environment, but also because mosquitoes in the affected countries have largely become resistant to it. These, it should be noted, are not microbes; they are complex animals, which, in less than 50 years, developed widespread resistance against the insecticide.

It is clear that the resistance mutations cannot have occurred in response to exposure to the drugs, except in a purely fortuitous fashion, should the drug, for example, have mutagenic properties. Any truly adaptive response would imply some sort of intentionality, which is strictly ruled out by our knowledge of molecular biology (see Chapter 7). No, the mutations were always there in some individuals or happened frequently, but only exceptionally were they of any use under natural conditions. Developing in the vicinity of a penicillium mold might be such an exception for a microbe. Barring such exceptions, it is clearly we who, by putting the drugs in the environment, have given the resistance mutations an opportunity to prove useful and thus to spread.

Many other natural phenomena illustrate the abundance of mutations. Mimicry is a good example. All are familiar with those insects that look for all the world like the leaf or branch they are sitting on, or with those fish that blend so well with the sea floor that an observer can hardly distinguish them. Such instances have often been brandished by adversaries of Darwinism as typical proofs of copying adaptations that cannot possibly be explained by natural selection acting on fortuitous genetic changes. Some instruction, it is claimed, must have been involved for such amazing similarities to arise. This need becomes less compelling once it is realized that natural mutations are so frequent and varied as to cover an immense spectrum of potentialities.

The many instances of artificial selection, which so impressed Darwin, are another case in point. Just think how, in a mere few thousand years, wolves have been led, by simple breeding methods, to produce fox terriers, shepherd dogs, St. Bernards, poodles, dachshunds, greyhounds, Pekingese, and all the other canine friends that humans surround themselves with. Who could have imagined, looking at the wild species, that they had it in them?

As another illustration of the enormous potential of chance mutations, it is a well-known fact among molecular biologists that almost any desired trait compatible with the cells' general organization can be elicited in a population of growing cells by sufficiently *stringent* culture conditions. Also impressive are the successes of the "carpet-bombing" technique of creating useful variants, mentioned in Chapter 7. Just throw your X rays or mutagenic substances indiscriminately, and you have a good chance of obtaining what you want, for example, a mold strain producing high yields of penicillin. There is even evidence that natural selection has retained a mechanism of this sort, whereby bacteria enhance the mutability of parts of their genome under stressful conditions, where survival of the population may depend on some rapid genetic readjustment.

In a slightly different vein, the immune system offers another example of the natural exploitation of genetic lavishness. The facts are well known. You are exposed to a foreign substance (antigen) carried, for example, by some pathogenic bacterium or virus, and, in a matter of two to four weeks, your body has built specific proteins (antibodies) that bind to the foreign substance by the kind of structural complementarity mentioned in Chapter 1, thereby serving to combat the invader. The same result is achieved preventively in vaccination, by exposure to an antigen-bearing organism previously rendered harmless by some appropriate treatment.

The remarkable thing about immunity is that it works with almost any foreign material of some chemical complexity. Hence the long-accepted assumption, seen as almost self-evident, that the foreign substance "instructs" the immune system to manufacture the complementary protein. In fact, this is not so. We now know that the immune system, in the course of its maturation, creates a wide array of genetically diverse cells programmed to make a correspondingly wide array of antibody proteins. All that exposure to the foreign antigen does is to trigger the multiplication of those few cells in the array that display the appropriate antibody on their surface and thereby advertise their ability to synthesize it. This is how the antigen boosts the production of the antibody. Thus, in the time it takes the immune system to reach maturity—some six months after birth—enough variation has been created (by a combinatorial program of genetic rearrangements) to cover billions of possibilities. The similarity with evolution is clear. At the start, there is wide variation, subsequently followed by selection.

Summing up this part of our analysis, we arrive at the seemingly paradoxical conclusion that, even though the mutations that allow evolutionary bifurcations are purely accidental occurrences, the chance factor is, nevertheless, largely abolished by the continually replenished plentifulness of the mutational field. In many cases, the decisive role is played by the environment, not the mutations. There are exceptions, no doubt, but, on the whole, what we witness of evolution in action tends to support this view. This still leaves chance an important role, however, by way of the environmental conditions that serve to screen genetic variants. Before addressing this question, we must look at another important aspect of genomes as targets of natural selection.

BODY-PLAN COMPLEXITY NARROWS DOWN
EVOLUTIONARY CHOICES

In a primitive organism, almost any mutation may be the start of a new evolutionary line. But, as complexity increases, the number of possible productive genetic changes decreases. To take a somewhat shaky simile, starting with a Ford model T, you have a wide range of evolutionary possibilities. Just look around while driving on a highway, preferably in Europe, where diversity is of worldwide origin, and you quickly get a sampling of the many different directions car evolution has taken. But with a Ferrari, the number of options is obviously restricted. There are thus *inner constraints* that limit further evolution.

In the case of genomes, these constraints are of two kinds. First, as already mentioned in the preceding chapter, the number of genes likely to be involved in evolution becomes progressively reduced to a limited set of "supergenes" as body-plan complexity increases. In addition, the number of possible changes of these genes that are compatible with the continuation of evolution does itself decrease as well, since the mutant form must be viable and capable of producing enough progeny under the prevailing conditions if it is to be retained by natural selection. The greater the commitment, the fewer the options satisfying this requirement. Many evolutionists have commented on this fact, which is often reflected in a progressive acceleration of evolutionary change. The remarkably rapid transformation of an ancestral chimpanzee into a full-blown human is a case in point. Apparently, once some key step was taken, there was little choice left but to continue ever faster in the same direction.

THE ROLE OF THE ENVIRONMENT

It has long been known from the fossil record that a world-wide catastrophe must have occurred some 65 million years ago, causing the extinction of the dinosaurs and of many other living species. In 1978, two American physicists, Luis Alvarez and his son Walter, found evidence that led them to suggest that the fall of a large asteroid was the phenomenon responsible for the cataclysm. This hypothesis has since been amply confirmed, and the impact site of the asteroid has even been located in what is now Chicxulub in the Yucatan Peninsula, in Mexico. This event is often cited as evidence of the far-reaching effects chance environmental circumstances may exert on biological evolution. But for a major blow from outer space, dinosaurs might still roam Earth, mammals might still be leading a modest and inconspicuous existence in the shadow of the big reptiles, and we would not be here to take notice of the fact.

Another often-quoted example is the formation, some six to seven million years ago, of the Great Rift Valley, which split a good part of East Africa. According to a theory proposed by the French paleoanthropologist Yves Coppens, codiscoverer, with the American Donald Johanson, of the famous skeleton known as Lucy,[16] this event may have played a major role in the emergence of the human species, by cutting off a group of apes from the forest and forcing them to adapt to the savannah, where bipedalism became a condition of survival and hands were freed to develop new skills. Had not Earth's crust cracked there and then, it is said, our ancestors might still be up in the trees.

Those are just two better-known instances of the no doubt numerous cases where some step in our evolutionary history has been crucially influenced by chance environmental conditions, providing incontrovertible proof, in the eyes of many leading evolutionists, of the unpredictability of evolution, in general, and of the *utter contingency of the human condition*, in particular. Long taken as irrefutably supported by modern science, this view is perhaps not as certain as it is often made to appear. Evolution proceeds in two directions, and these are affected in very unequal fashion by the environment.

The Two Directions of Evolution

Every evolutionary step depends, as we have seen, on a genetic change that, in the case of a multicellular organism, has affected a germ cell that will later be involved in the generation of a fertilized egg. In order to play a role in evolution, this change must be reflected in a modification of the organism's developmental program and it must be transmissible, that is, the modified individual must be able to produce progeny. Whether the change will eventually give rise to a new branch depends on the environment, including the necessity for reproductive isolation mentioned in the preceding chapter. What will happen also depends on how significantly the body plan is affected by the genetic modification. Here is where the distinction between the two directions of evolution, which I like to call horizontal and vertical—roughly equivalent to what is often termed microevolution and macroevolution—comes into play.

HORIZONTAL EVOLUTION IS THE REALM OF CONTINGENCY

Most often, genetic changes are minor and do not basically modify the body plan. They are, so to speak, variations on the same theme. More than 750,000 insect species are known and several million may remain to be discovered. Even though they may be as different as a butterfly, a scarab, or a praying mantis, they are all built according to the "insect" blueprint. This type of evolution, which is the one I designate as horizontal, is largely responsible for *biodiversity*, that is, the extraordinary variety of forms evolution has created on given models, be they of a grass, insect, fish, or mammal.

In horizontal evolution, chance plays the star role. It enjoys an almost free rein, flitting over a multitude of special environments where small isolated groups are allowed to develop certain peculiar traits closely connected to the local scene. The cases of mimicry alluded to earlier are typical examples. Darwin's famous finches are another. In the course of his voyage to the Galapagos Islands, Darwin observed that, on each island, finches had differently shaped beaks, adapted to the kind of food the island offered. These various species, Darwin reasoned, must descend from the same ancestral species, which, in the reproductive isolation offered by each island, evolved separately according to the local conditions.

Innumerable such examples could be summoned. Horizontal evolution is indeed the realm of contingency, largely shaped by the accidents that create, fortuitously and unpredictably, conditions favorable to the expression of one among the many mutations it is continually presented with. (Remember drug resistance.) This is how each type of organizational pattern has produced, by progressive horizontal ramification, thousands or more different species adapted to a wide variety of ecological niches. Even here, however, as I shall point out later, certain privileged directions, manifested by convergent evolution, may be discerned.

VERTICAL EVOLUTION IS CHANNELLED

But evolution has not just composed variations on the same theme; it has also created new themes. It has done so through modulation from existing themes, by way of genetic modifications that were both rarer and more exacting than those involved in horizontal evolution and that have led to much more important, sometimes even spectacular, rearrangements of the body plan, while all the time remaining compatible with the survival and reproductive success of the new forms.

In many cases—not all, since examples of regressive evolution are also known—this type of evolution, which I term vertical, involves an increase in *complexity*. I use this word intentionally, rather than "progress," which, because it implies a judgment of value, understandably upsets a number of contemporary thinkers. Even the word "complexity" is the object of many learned philosophical discussions. For my part, I shall stick to the intuitive meaning most of us attach to it, as opposite of simple. And I shall take it as self-evident that vertical evolution, that is, the one that involves significant changes in body plan, has, with time, produced living beings of increasing complexity. The two directions I singled out as privileged in the preceding chapter, leading to increasingly efficient reproductive means in both plants and animals and to increasingly intricate polyneuronal networks in the animal line, clearly illustrate the course of vertical evolution towards increasing complexity. Or, to take an even less subjective piece of evidence, the fact that evolution has, in both kingdoms, produced an increasing number of differentiated cell types should demonstrate to almost anyone's satisfaction the reality of vertical evolution toward increasing complexity.

Vertical evolution, thus defined, is subject to much more stringent inner constraints than horizontal evolution. It is intuitively evident that transforming a body plan—say, from a fish to an amphibian—must be more difficult than just modifying it in some trivial fashion—such as turning an ancestral fish into a sole or a mackerel—especially because each intermediary form has to be viable and able to reproduce. Going back to our simile of car evolution, it is easier to convert a model T into a Jeep or a Ferrari than into a helicopter, especially if the conversion is to occur through a series of intermediary forms each of which is fit for transporting passengers with reasonable efficiency.

In concrete terms, this means that there are fewer genetic options that can propel evolution vertically than can do so horizontally. The difficulty increases—and, thus, the number of options decreases—as complexity increases. This, also, is evident. A simpler body plan is likely to be more flexible and more tolerant to change than a more complicated one.

All this goes to show that there is, written into the very structure of genomes, more inevitability in vertical than in horizontal evolution, and that the degree of inevitability increases with the progression of vertical evolution. This seems particularly true for the evolution of the nervous system of animals toward increasing complexity, which is almost independent of the environment, since it is hard to imagine an environment in which possession of a better performing brain would not be an advantage for an animal. In this sense, progression toward complexity may be viewed as obligatory in so far as it is genetically and environmentally possible.

This point is very relevant to the origin of humankind. I mentioned earlier two major fortuitous events—the fall of the asteroid that killed off the dinosaurs and the opening of the Great Rift Valley, believed to have isolated our ancestors from the forest—that may have played a key role in human evolution and are often alleged as proof of the contingent character of humanity. The argument is impressive, but not irrefutable. It is quite possible—some might even say probable—that environmental vagaries have done no more than determine the moment when an inevitable evolutionary development actually took place. Perhaps mammals were bound to supplant dinosaurs at some stage for reasons linked to the intrinsic properties of the two types of animals, and the asteroid only precipitated an event that would have occurred sooner or later. Similarly, had not the Great Rift Valley created conditions suitable for hominization, some other geographical upheaval would have done so at a later date. The same could be true of many of the other circumstances that have allowed a decisive step in the long, vertical progression that led to the human species.

RERUNNING THE TAPE

"Wind back the tape of life . . . ; let it play again from an identical starting point, and the chance becomes vanishingly small that anything like human intelligence would grace the replay." This oft-quoted passage from Stephen Jay Gould's best-selling *Wonderful Life*[17] has become for much of the educated public the dire but unassailable message from modern biology, underlining the utter contingency of the human condition.

The British paleontologist Simon Conway Morris is not impressed with the message. An expert, as was Gould, on the Burgess Shale, a geological site in the Canadian Rockies about 530 million years old and rich in fossils of bizarre animals (which forms the main topic of Gould's book), Conway Morris devotes much of a recent work[18] to demonstrating that the Burgess Shale fauna is by far not as weird as is claimed by Gould. At the end of his book, Conway Morris addresses the "rerunning-the-tape" image. He considers the argument both trivial, in that it simply repeats what has always been known of the contingency of historical events, and false, to the extent that it applies to the main directions of evolution. "In fact," he writes, "the real range of possibilities and hence the expected end results appear to be much more restricted. . . . Within certain limits the outcome of evolutionary processes might be rather predictable."[19] In addition to some of the constraints already mentioned above, Conway Morris lists, in support of his contention, the increasing number of known instances, some truly astonishing, of *convergent evolution*.

Animals that have been separated and left to evolve independently for up to several hundred million years are found to develop into extraordinarily similar types. Anteaters have the same intricate specializations in North America, South America, Africa, and Oceania. So have felines dependent on hunting or herbivores built for speed. It is the case also with underground mammals. According to a world authority on the topic, the Israeli biologist Eviatar Nevo, the 285 known species of these animals share in astonishing fashion *"regression, progression and global convergence."*[20]

What these and many other similar cases of convergent evolution tell us is, first, that there are not so many solutions to, for example, the problem of surviving on ants and, especially and more surprisingly, that, under sufficient selective pressure, the same solution will be found time and again by different species. Even trivial aspects of horizontal evolution may apparently be subject to this kind of channelling. "Rerunning the tape"—an image we should be careful not to take too literally, in any case—no doubt would not produce exactly the same script but might well come sufficiently near the original story to make it perfectly recognizable.

THE PRICE OF A BRAIN

Watch a mare giving birth. The animal shows signs of only minor discomfort. After delivery, the newborn foal staggers to its feet in a matter of minutes and lurches toward its mother to partake of its first meal. Now move to the delivery room in a hospital and watch the human version of the same scene. It takes the mother hours of excruciating pain—with a number of attending risks that often proved fatal before the advent of modern medicine—to give birth to a baby that is entirely helpless and will need many months to reach a stage where it can match the newborn foal's autonomy. Faced with these two contrasting scenes, even the most confirmed Darwinian may wonder how natural selection ever retained a mode of giving birth as hazardous as the human one. The key to the riddle, at least as I see it, is: a *better brain*.

In humans, development of the brain takes place inside the womb until the size of the head has reached the utmost limit compatible with passage through the birth canal. Even then, the brain is still at a very immature stage and continues to undergo outside the womb a developmental process that, in other mammals, is largely completed *in utero*. It is thanks to this delayed development—the technical term is "neoteny"—that the human species has been able to achieve its unique brain size and to acquire the extraordinary accompanying mental attributes. No other fact could illustrate in a more striking fashion the force of the selective pressures favoring vertical evolution towards increasing polyneuronal complexity.

THE HUMAN SPECIES OCCUPIES THE TOP OF THE TREE OF LIFE

Since Haeckel, the course of evolution is, for excellent reasons, represented by a tree. For the German zoologist-philosopher and for the other early disciples of Darwin, imbued as they were with Victorian triumphalism, this tree rose majestically towards a summit that nobody doubted was dominated by humankind, the uncontested master of creation.

Today, the point of view has changed. It has become politically correct to put the emphasis on the canopy of the tree and on the millions of terminal twigs that compose it. The human species, it is pointed out, is no more than one of those twigs, on par with plague bacilli, amoebae, orchids, scorpions, baboons, and all the other species that form the thin living pellicle that surrounds Earth. Like the others, the human twig is the outcome of some four billion years of evolution, the result of thousands of accidental mutations screened by natural selection according to the caprices of environmental conditions. Only our insolent vanity gives the twig a special significance, which is justified by no objective reason. On the contrary, if any discrimination is to be made, it should, as mentioned in Chapter 8, favor the much more ancient and sturdier bacteria.

Calling on science to denigrate the human species is part of the so-called deconstructionist, post-modernist trend, which blends science, philosophy, sociology, and politics into an ideological mixture inspired by a negativistic, relativistic vision of knowledge, which goes so far as to deny the very existence of objective reality. I do not feel qualified to discuss this topic, which has been addressed by many contemporary authors. Remaining at the level of my competence, I shall limit myself to pointing out that the image proposed for the tree of life is a perversion of the scientific facts on which it is allegedly based.

Like all trees, the tree of life has grown simultaneously in two directions: vertical and horizontal. It is true that if one looks only at the canopy, one is first struck by the biodiversity created by horizontal evolution. But this is only a very superficial vision. The early naturalists already distinguished similarities in the diversity of life and classified it in a series of hierarchical divisions. With the advances of paleontology and, more recently, of molecular biology, this hierarchy has been found to correspond to a historical reality and to reflect a series of stages in the growth of the tree. The lower branches, which detached earliest from the trunk, end up in the simplest forms of life. As we move upward, and the proportion of evolutionary time taken by vertical growth increases, terminal twigs bear increasingly complex forms of life. Thus, even the tree's canopy, when properly examined, already appears as hierarchically organized.

The inner structure of the tree is not directly visible but has been revealed by modern science. As could be suspected, the visible terminal twigs are derived, by horizontal ramification, from hidden master branches that have detached from the trunk in the hierarchical order dimly revealed by the canopy. Many key bifurcations in this development have now been recognized, and the properties of the organisms that occupied them are beginning to be appreciated.

This fact calls for a remark. In everyday language, it is often said that birds descend from dinosaurs, reptiles from fish, or humans from monkeys. Used for simplicity's sake, such expressions may be quite misleading if one doesn't pay attention. Today's reptiles were not born from today's fish. They share with today's fish a common ancestor that lived some 400 million years ago. This ancestor was the starting point of a bifurcation of which one branch continued to diversify into fish, whereas the other evolved progressively up to another critical bifurcation leading, on one hand, to amphibians and, on the other, to reptiles. These long-gone transitional forms make up the trunk of the tree of life. They could have been very different from their present-day descendants. All we can do is try to reconstruct them with the information provided by the descendants and by fossils and vestigial organs.

The tree of life manifestly grows in two directions, vertically toward complexity and horizontally toward diversity. It has a top, albeit in the form of a thin terminal twig among millions of others. It is obvious that the human twig occupies this top, at least if the brain is adopted as the criterion of complexity. Considering all that the human brain has produced, one must have a distorted set of values to refuse to accept this criterion.

If our situation on top of the tree of life should leave little doubt, this is no reason for bragging. All we know of the history of life makes it likely that our eminent position is only temporary. It was occupied three million years ago by a young female called Lucy[21] and, three million years earlier, by the last common ancestor we share with chimpanzees. What form of life will occupy it in the future is anybody's guess. The astronomers tell us that Earth should be able to sustain life for at least 1.5 billion years, possibly up to 5.0 billion years.[22] If the tree of life goes on growing vertically, it could reach more than twice its present height. Our imagination is totally incapable of foreseeing what kind of being may emerge from such a process. Note that this development does not have to happen through further growth of the human twig. There is plenty of time for a more promising vertical line to start from another twig while our own twig withers. This thought should inspire a solid dose of humility, together with a reappraisal of some our most cherished notions concerning our significance.

13. Becoming Human

T HE HUMAN TWIG DETACHED FROM THE PRIMATE BRANCH ABOUT
six million years ago. The last ancestor we share with another
existing primate probably resembled a chimpanzee, our nearest
extant relative. Evolution from this ancestor favored erect walking, manual skill, and a number of other traits that are clearly evident when humans and apes are compared. Most impressive is the development of the
brain, which took place at a staggering speed. After having taken some
600 million years to reach a volume on the order of 450 cm^3 in our
simian ancestors, the size of the hominoid brain went through an astonishingly rapid phase of expansion, virtually jumping to three times this
value in little more than two million years. On the evolutionary time
scale, such a rate of change is no less than dazzling. It illustrates in a
remarkable way how fast vertical evolution may rush when given the
opportunity. Note, however, that the rate of change still would have been
hardly noticeable on a human time scale, amounting, on average, to a
volume increase of about ten cubic millimeters—no more than a pin's
head—per generation.

Paralleling this increase in brain volume and, no doubt, directly related
to it, there is the remarkable enrichment of mental processes that distinguishes human beings from all other animals, including their closest primate cousins. In the past, this distinction posed no difficulty, as it was
viewed as an authentic dichotomy, a difference in kind, readily explainable
by the notion of a separate creation of the human race. The discovery
that a continuous evolutionary process links humans to the other primates has radically upset the classical view of humanness, creating problems that are still far from solved. It is generally recognized that human

mental processes are rooted in primitive phenomena that already existed in animals long before the final hominization leap of evolution and that, like every other evolutionary attribute, these processes owe their development to a succession of accidental genetic modifications screened by natural selection. But the details of this history are poorly understood and remain the subject of considerable debate.

The Roots of Mental Life

The old distinction that grants a psychic life exclusively to human beings is no longer accepted. Most ethologists admit that consciousness is experienced by some animals. Some experts even go so far as to endow invertebrates, such as bees, with a rudimentary consciousness. We need not follow them to that point to attribute such a faculty at least to mammals. This is taken as self-evident by most of those who have ever lived with a dog or cat. As to primates, there is now a long history of observations in the wild and in the laboratory clearly indicating that they enjoy some sort of mental life crudely similar to what goes on in the human brain.

THE ORIGINS OF CONSCIOUSNESS

In the opinion of many experts, consciousness first manifested itself by way of feelings associated with certain states or actions. Thus, pleasant feelings would, from a given degree of cerebral development, which need not have been very high, have accompanied physiological activities such as feeding, excreting, or mating. Such feelings would have served to consolidate the underlying nervous mechanisms, with as corollary that their absence would, by the painful sense of deprivation it generated, have stimulated behaviors aiming at satisfying—the term is revealing—the needs in question. In a different range, fright would have been associated with flight from danger, and fury with attack against an enemy or rival. One readily sees how natural selection would have favored such affective complements to physiology, inasmuch as they reinforced behaviors useful for survival and reproduction.

A crucial step in hominization has been the acquisition of self-awareness. This trait seems to have been acquired late. A chimpanzee seeing itself in a mirror first tries, with signs of growing perplexity, followed by irritation, to socialize with the reflection. The animal eventually ends up recognizing the image as its own, by what appears as a fairly intricate mental process based on correlation of the image's gestures with its own. But this is as far as the awareness of self seems to go in animals. They do not have our sense of personhood nor our perception of time as a key dimension in which our personal history unfolds. They seem to be completely ignorant of their own impending death and its inevitability.

THE ENGINEERS PRECEDED THE THINKERS

Tools most likely led the way to rational thought. Many are familiar with the oft-repeated story told by the British primatologist Jane Goodall, who has spent many years in the company of chimpanzees in the Gombe Reserve in Tanzania. It tells of a chimpanzee that plucks a small branch, strips it of leaves and lateral twigs, thrusts it into a termite nest, waits a moment, and subsequently draws it back to feast on the insects that cling to it. Many examples of tool use, not only by primates but also by other animals, have been observed. To open shellfish, sea otters use a flat stone they place on their belly as they swim on their back and against which they bang the molluscs until these break. Birds are known that crack the shells of eggs of other species by dropping on them a stone they hold in their beak. Even insects, like some ants, exhibit behaviors reminiscent of tool use.

Such manifestations have long been attributed to instinctive programs, imprinted into the genomes by natural selection and executed blindly, without conscious design or control. The fact that animals were denied any sort of consciousness was evidently not foreign to this view. Now that ideas are changing on this subject, greater significance is being attached to observations indicating that the gestures are not as stereotyped as was believed and are adapted to the prevailing circumstances. It is increasingly suspected that there may be behind many a useful gesture some sketchy mental representation of its function. When it comes to the termite-fishing chimpanzee, it is difficult not to see in the animal's behavior the accomplishment of a project conceived in anticipation of a given result.

This faculty continued to develop in the early days of hominization. The first chipped stones obviously made for a purpose date back more than two million years. They were manufactured by beings that would not be recognized as human today. Anthropology museums all over the world display evidence of the exceedingly slow process, covering many hundreds of millennia, whereby these primitive tools were progressively improved and adapted to different uses, changing from grossly cut pebbles to the elaborately shaped bifaces, axes, picks, scrapers, and other specialized tools and weapons of the end of the Paleolithic period.[1] Could it be that this history took such a long time because it was conditioned by brain expansion and the related development of intelligence?

It is widely believed that toolmaking played a key role in the development of the human intellect, by way of an evolutionary to-and-fro interchange between hands and brain. Freed by bipedal walking, the hands came to serve to a greater extent for prehension. The gestures thus accomplished often being beneficial to the preservation and propagation of the species, any genetic modification tending to make the use of hands more efficient had a good chance of being retained by natural selection. Improvements acquired in this way could have been anatomical, such as the opposing thumb. But they also, and perhaps most frequently, could have affected the cerebral mechanisms governing the gestures. Many experts see this as being, at the start, the main source of the selective pressure that advantaged expansion of the human brain. *Homo sapiens*, knowing man, is issued from *Homo habilis*, handy man.

It is likely that social organization in distinct, closely knit groups also played an important role in this development. Prehumans probably formed small roaming bands, subsisting by hunting and gathering and mostly keeping away from other, similar bands. Anything that favored communication and collaboration among members of a group could only contribute to the group's evolutionary success. Inbreeding linked to reproductive isolation probably was more useful than harmful at this stage. Even though many groups may have succumbed to unfavorable genetic combinations, those in which an individual benefitted from a favorable mutation could, in a few generations, have this mutation spread among all members of the group. Diffusion of the favorable gene was further advantaged to the extent that the mutation contributed to reproductive success.

Thus, probably, did the larval form of intelligence present in our simian ancestors develop slowly into a more complex faculty, harnessed mainly to the practice of manual activities and to the preservation of the social fabric. One readily understands that possession of such a faculty may have constituted a selective advantage contributing to the proliferation of the individuals so endowed. But this advantage would have remained limited if each individual could rely only on the genetically imprinted faculty to foresee, design, and execute. What truly made the difference was the fact that acquisitions made with the help of this faculty could themselves also be transmitted from generation to generation. Even though such transmission could not take place by the hereditary mechanism imagined by Lamarck, it could occur by *communication*. As we shall see later, this phenomenon launched the process known as *cultural evolution*.

The pace of cultural evolution started accelerating in remarkable fashion some 50,000 years ago. All of a sudden, wood, bone, and horn were added to stone as materials for manufacturing tools and weapons. Humans started constructing shelters and ships. They began to bury their dead, carve figurines, decorate the walls of caves with extraordinarily vivid paintings, make the first jewels and ornaments. A whole, complex civilization thus was built in the space of a few millennia, leading finally, some 10,000 years ago, to the first sedentary communities, which replaced gathering with agriculture, and hunting with domestication and breeding. During that time, the structure of the human brain probably did not change very much. Progress resulted from the fact that each generation retained the heritage of the past and added its own contribution to it. According to many experts, the advances of *language* are mainly behind this sudden acceleration of cultural evolution.

THE DEVELOPMENT OF LANGUAGE

There is little doubt that animals communicate with each other by a variety of means, which involve every one of the senses. Even insects can do so, as illustrated by the famous dance, elucidated by the Austrian ethologist Karl von Frisch, whereby bees that have discovered a rich source of nectar communicate its location to other members of the hive. In the 1930s, it took only a few years for the tits of a good part of England to be informed of the discovery, made by one of them, that a rich meal of cream could be had by tearing open the tin foil capsule of the bottles left on doorsteps by milkmen. Vervet monkeys have watchers that use different sounds to warn their congeners of the presence of a leopard, a snake, or an eagle.

Learning plays an important role in animal societies. Many examples are known of animal behaviors of which only the general scheme is transmitted genetically, the details being subsequently filled in by imitative learning. The construction of nests and other abodes, hunting techniques, bird songs, and various social rituals are examples. All this forms a "lore," the stirrings of a culture, transmitted across generations by nongenetic pathways. The 17 June 1999 issue of the magazine *Nature* contains an article surveying 65 behavioral traits in seven wild chimpanzee populations located in different parts of Africa. The paper's conclusion is unequivocal: each population had a distinctive set of shared patterns, almost certainly of nongenetic origin, justifying the paper's title: *Cultures in chimpanzees.* A little more than one year later, whale songs made headlines, under the title "cultural evolution."[2]

Some primates even appear capable of *symbolic communication*. A chimpanzee, named Washoe, became famous in the 1970s for having learned to distinguish more than 100 signs, in the laboratory of an American couple, Allen and Beatrice Gardner. The record belongs to Kanzi, a bonobo, that, under the guidance of another American, Sue Savage-Rumbaugh, learned to converse with the help of a keyboard. According to a recent compilation, Kanzi comprehends at least 500 words, and the list is still growing.[3] Among other mammals, dolphins likewise seem to possess relatively elaborate means of communication, which go together with a cerebral development comparable to that of nonhuman primates.

There is, however, a limit to the extent of animal communication. Part of it may be anatomical. In most nonhuman mammals, the larynx is close to the mouth, with as a consequence a strict limitation in the variety of sounds that can be emitted. It is possible that animals don't talk because they are physically incapable of doing so and have, for this reason, never acquired the corresponding cerebral faculties. The human larynx is situated further from the mouth and is believed to owe its capacity for articulated speech to this anatomical trait, which is actually acquired during the first ages of life. The newborn baby can utter only indistinct sounds and learns to speak as its larynx moves down.

In the eyes of many observers, this is an example of the "recapitulation law" (which suffers numerous exceptions), proposed by Haeckel in the nineteenth century (see Chapter 7), according to which steps in development reproduce the main steps in evolution. Thus, descent of the larynx, observable today in a late developmental phase, would be the evolutionary phenomenon that made articulate speech possible and, through this, offered the necessary physical substratum for the cerebral development from which language arose. According to the American linguist Philip Lieberman, who has made it a specialty to reconstruct the anatomy of the phonetic apparatus from moldings of fossil skulls, this trait exists only in the ultimate hominization twig, *Homo sapiens sapiens*, which dates back less than 200,000 years. It was lacking even in Neanderthal man, *Homo sapiens neandertalensis*, which became extinct a bare 35,000 years ago and, according to Lieberman, was unable to speak. This opinion, however, is not unanimously accepted.

THE BIRTH OF SCIENCE

Probably retained originally by natural selection as a means of social communication, language simultaneously made possible, as a sort of evolutionary fringe benefit, the kind of internal soliloquy that underlies rational thought, leading in turn to the first attempts at understanding the world. Mythologies, religions, and philosophies were developed in turn in this quest for explanations, to be followed, at a late stage and not everywhere, by the scientific enterprise.

Historically, the search for knowledge has long been preceded by purely practical preoccupations. Many technical advances were first accomplished empirically, without benefit of prior knowledge. Their inventors were tinkerers of genius who, by trial and error and profiting from past experience for making improvements, fashioned tools, manufactured weapons, constructed machines, built cities, raised fortresses, found medicines, exploited natural energy sources, conquered the seas, in short, created the first technically evolved civilizations. Success often came before understanding. Thermodynamics, for instance, was developed to explain the conversion of heat into work long after steam engines had started pumping water, propelling boats, and pulling trains.

From these empirical, utilitarian roots, there was born, in the course of time, a new form of exploration of the unknown, which has become modern science. Motivated, like the philosophies of the past, by the sole desire to understand, the scientific endeavor has proved immensely more powerful than the philosophies, thanks to a strategy based on observation and experiment, admittedly guided by reflection and reasoning but independent of any dogma or preconceived idea (at least in principle). Interestingly, this conversion has happened almost exclusively in the civilizations that have developed around the Mediterranean basin. The Chinese, for example, who produced one of the most advanced technological cultures, never made the step to scientific understanding.

The development of science, in turn, has spawned a new form of technological research, inspired and illuminated by previous knowledge. Basic research has led to applications of immense importance, which would have been utterly unthinkable before acquisition of the knowledge from which they issued. The exploitation of atomic energy and biotechnologies are spectacular examples of such developments. But there are many others, notably in medicine and informatics. They illustrate a new means of mastering and manipulating nature, based on understanding. One starts by investigating how things function and then goes on taking advantage of the acquired knowledge to rationally conceive a practical application.

Even more important than such applications are the contributions of science to human culture. Irrespective of its value as a source of new, beneficial technologies, basic research has proved an inestimable generator of *knowledge* and *understanding*. It has, largely in the space of a single century, elucidated the nature of matter, established the composition and history of the universe, unravelled the most intimate biological mechanisms, uncovered the origin and evolution of life on Earth, traced the advent of humankind, finally to approach the functioning of its own motor, the human brain. Those are historic events of immense importance. Not only do we owe to them means of unprecedented power for shaping the future; they have also given us a totally transformed vision of the world and of our nature.

THE ORIGIN OF ARTISTIC EMOTION

Words do not serve merely as vehicles of information; the manner in which they are assembled communicates more subtle messages, directed at faculties other than intellect. When we read *"et rose elle a vécu ce que vivent les roses, l'espace d'un matin"* (and rose she lived what roses live, the space of a morning), an excerpt from the *Consolation à Monsieur du Périer*, written by the French poet François de Malherbe, who straddles the sixteenth and seventeenth centuries, on the occasion of the death of Mr. du Périer's daughter Rose, our brain catches the meaning of the sentence, but it does not analyze it coldly, wondering what variety of rose is being referred to or from which botany textbook the author has drawn his information on the ephemeral character of those flowers. No, something undefinable in us resonates with the nostalgic rhythm of the words, shares Mr. du Périer's sorrow, and finds, together with him, a little of the solace the poet is trying to impart. The poetic message goes through our intellect, but it is addressed to another part of our brain, which we sometimes call our heart or our "guts."

This kind of emotional experience occurs in response to all other forms of art. Whether it be music, painting, sculpture, or any other work of art, there is an intellectual aspect—the work's construction—accessible mostly to the expert, as is physics to the scientist. But that is not the main thing. What counts is the emotion the work elicits. Varying according to individual personalities, this emotion is not closely reasoned; it is perceived to some extent without the mediation of intellect and gives the impression of bringing us in direct contact with some hidden facet of reality. It is the artist's gift to be particularly sensitive to this facet of reality and to have the talent to reveal it. By way of their works, some kind of resonance is created between their emotion and ours, intensifying our perception. Our emotion is not necessarily the same as that of the artists; but it is of the same essence.

This communication—or, better said, *communion*—is particularly subtle for arts that require an interpreter, especially those, such as theater, opera, ballet, or cinema, in which several artistic forms are combined. Even pure music involves a visual aspect that is far from negligible. A given piece does not make the same impression heard live or from a record. I have always been impressed with the beauty that illuminates the faces, even those most disadvantaged by nature, of the members of an orchestra or choir as soon as they start performing. Nobody has captured that transfiguration better than the Flemish painters, the brothers Hubert and Jan van Eyck, in the musician angels of their celebrated polyptic, the *Adoration of the Lamb*, which hangs in the Sint Baafs cathedral in Ghent. Look at those angels and you hear the music of the heavens.

Artistic emotion has probably grown from those primordial feelings, associated with the awakening of consciousness, that accompany the accomplishment of certain acts or functions. From feeling would have arisen, by way of memory, a mental representation of the object that causes pleasure or pain, leading in turn to imagination, a prerequisite both of intellectual speculation and of artistic expression.

If feelings and emotions are probably, in more or less evolved forms, prerogatives of many animal species, art appears to be specifically human. Only birds seem to display by certain manifestations—their songs, for example, or the bowers the males of certain species decorate for their amorous parades—behaviors to which we would be tempted anthropomorphically to attribute an esthetic dimension. But one may wonder whether the emotions that accompany those sexual strategies, retained by natural selection for their reproductive usefulness, have anything in common with what we call artistic emotion.

Be that as it may, art, as a form of expression and communication, is a universal characteristic of humankind. There is no people, however primitive, that has not accumulated a cultural heritage of legends, music, dances, and pictorial and plastic representations. Art is ancient, as proven by prehistoric cave paintings, whose astonishing degree of maturity is recognized by experts. Whereas artistic feeling is universal, the forms under which it is expressed vary according to populations, epochs, and individuals. Those forms seem to be less the expression of absolute canons than the result of historical contingencies fixed by tradition and education.

At first, the function of artistic production was probably mainly social, linked, for example, to myths, symbols, or rites constructed around the conjuring of divinities, or to festivities, celebrations, rejoicings, or homages rendered to a leader, the dead, or a loved one. Art also came to play a historical role, notably by reproducing the features of famous men or women and by commemorating victorious battles or important political events. Collective entertainment has also often served as an occasion for artistic creation. Finally, art is increasingly used as a medium for driving through, by means of appropriate emotional reinforcements, a given social, political, economic, or commercial message.

In the course of time, however, artistic expression has progressively freed itself from its functional aspects—a little like science has freed itself from technology—to become to some extent a gratuitous activity. There has appeared an "art for art's sake," as there is a "science for science's sake" and a "mathematics for mathematics's sake." The appreciation of art has followed a similar trend. When we go to a concert or visit a museum or gallery, our main preoccupation is with the quality of the emotion evoked in us by the music we hear or by the works we see. We each react differently but we share an undefinable something, a feeling of somehow communing with an aspect of reality different from that accessible to the intellect.

THE SOURCES OF MORALITY

Many scientists believe that ethics is a product of natural selection, which is taken to have retained social behaviors favorable to the evolutionary success of groups. Animal societies show many examples of cohesion based on the instinctive submission to what appear like unwritten laws. The primitive groups ancestral to the human species no doubt had an organization of that kind, which, with the development of cerebral faculties, progressively transformed into the institution of, and respect for, explicit legislations. The societies that gave themselves laws and applied them proved better able to survive and proliferate than those left to anarchy and savage competition among members.

This conception has been widened by the Harvard biologist Edward Wilson, under the name of *sociobiology*, to encompass the entire human social fabric. According to Wilson, who has summed up his views in a major work, *Consilience*,[4] our whole system of values, including the beliefs, virtues, and rules related to them, is purely a product of evolutionary expediency. The system exists simply because it happened to be useful to the evolutionary success of the groups that entertained it.

Sociobiology has been vigorously opposed for various reasons by many philosophers and social scientists. Some see in it vestiges of social Darwinism, the doctrine defended, notably, by the nineteenth-century British philosopher Herbert Spencer, to justify, on the strength of Darwin's theory, the excesses of economic *laissez-faire*. In the view of others, sociobiology exaggerates the role of genetic determinism, to the detriment of environmental influences, and encourages racial and social discriminations. Finally, the thesis of a natural origin of ethics is obviously not accepted by those who believe that moral rules were edicted by God when He gave Moses the Tables of the Law on top of Mount Sinai.

Leaving aside such ideologically charged polemics, two simple reflections come to mind. First, it seems hardly disputable that societies subject to laws had greater chances of success than lawless ones. On the other hand, comparative anthropology clearly shows that laws vary according to peoples and epochs. Natural selection thus played a role; but what it has encouraged is the existence of laws, not necessarily the details of their content.

Whatever the origin of our ethical behavior, there are good reasons for believing that, with the development of the brain, morality has evolved progressively from a purely pragmatic and utilitarian form to a more abstract conception of good and bad. Most civilizations make a distinction between legislations, dictated by considerations of expediency, and ethical rules, based on absolute values. These remain to some extent arbitrary, as is shown, for example, by the many debates on bioethics. But the very distinction between good and bad seems to be rooted deeply in human nature.

ALTRUISM AND LOVE

Even the most powerful human sentiment does not escape an explanation calling on natural selection. Maternal love, mate bonding, altruism have their equivalent in numerous animal species. It has been pointed out that such behaviors favor the protection and multiplication of genomes and may, for this reason, have been selected in a purely passive fashion.

This explanation is plausible for the instinct that drives a female to go so far as to sacrifice her life in defense of her young. From the point of view of evolutionary success, several young lives are worth more than a single old one. Even a single young life is likely to produce more progeny than an older one, provided, of course, it has acquired a sufficient degree of independence. The solidity of the bond between sexual partners, although it varies considerably from one animal species to another, generally parallels, as predicted by the theory, the importance of the joint roles of the two sexes in reproductive success. Birds are a characteristic example. Finally, even altruistic behavior—the sacrifice of one member of a group for the group's benefit—is explainable by natural selection. This has been shown by the late British biologist William Hamilton in his theory of kin selection, which evaluates the evolutionary benefit of the sacrifice as a function of the degree of kinship between the "altruist" and the other members of the group. There is benefit if the loss of the altruist's genes allows a greater number of the same genes to be saved in the group.

All this, we are told, is explained by the fact that individuals genetically disposed to protect their young at the risk of their lives, to remain united in a manner that favors the welfare of their offspring, or to sacrifice themselves under circumstances such that a danger threatening their kin is lessened have a greater chance of propagating their genes, and thus their behaviors, than those devoid of those genetic characteristics. Those are seen as mathematical truths. Whether feelings are associated with such behaviors may be guessed from the attitudes of the animals, but in a purely anthropomorphic framework; we don't know what goes on in their minds.

The naturalistic explanation of altruism is convincing. But thence to reduce to the mere play of the evolutionary lottery the flame that burns in the hearts of lovers, the tenderness of a mother for her child, the complicity between two old people contemplating, hand in hand, a lifetime spent together; to bring down to a purely utilitarian function the sentiment that has inspired so many poets, writers, musicians, and artists, motivated so many heroic acts, and engendered so many bitter rivalries and conflicts, even between nations; there cannot be many prepared to take this step. Love transports, transfigures, gives a feeling of participating in a sort of cosmic rapture, to the point of sometimes blinding the senses and reason. As with other mental manifestations, one has the impression that the development of the human brain has drawn love out of its primitive shell and allowed it to blossom in a more subtle sphere, a sort of "love for love's sake," so to speak.

INVENTION OR DISCOVERY?

The cardinal notions truth, beauty, goodness, and love have, for several millennia, fed the ponderings and discussions of philosophers. Beginning with the ancient Greeks, a central question has been whether, as Plato proposed in his famous myth of the cave, those notions exist outside of us, waiting to be *discovered*. Or are they products of human *invention*, creations of the growing complexity of our brain?

Modern science has contributed valuable new elements to this debate by revealing that our lofty ideals have humble roots. Each of the philosophers' abstractions, science tells us, seems to originate from ancestral phenomena that began to manifest themselves long before the advent of the human species and developed thanks to selective advantages linked with their emergence. Our search for truth may stem from simple, manual acts; our aspiration after beauty from lowly visceral feelings; our pursuit of goodness from practical social necessities; and our yearning for love from primitive survival strategies. In each case, the traits have developed through a long succession of genetic changes, most often affecting the brain, that were retained by natural selection because they contributed to the survival and reproduction of the individuals concerned and their kin. The success of these notions is purely pragmatic and in no way attests to their authenticity.

For a number of contemporary thinkers, led by Wilson, the father of sociobiology, the message is clear. There is nothing else. The whole system of values we associate with the word civilization is built on strictly utilitarian foundations, unrelated to any external reality that might transcend us. The human sciences, the philosophies, and the religions have no other recourse than to face this evidence, accept it, and adapt to it.

Such a conversion is not easy. Wilson himself acknowledges this. How can we, in the light of cold reason, erase what millions of years of selection have imprinted in the very depths of our nature? How can we replace the whole sociocultural heritage of the last few millennia with an impersonal message based on premises that are inaccessible to the majority and contested, in addition, by part of the minority that understands them? Is such an upheaval truly necessary?

Perhaps not. Take the contributions of science. Surely, they are not inventions; they are discoveries, subject, to be sure, to amendment or, even, abandonment, if new facts demand it; but they are, by virtue of this very caveat, at least steps in the direction of discovery. Remarkably, the picture of reality emerging from this effort is totally different from the familiar image that supports our everyday life; so different as to require for its description a highly specialized mathematical language mastered by only a handful of experts. Yet, few of us will deny that the new picture emerging from modern science is bringing us closer to what may be called *ultimate reality*. In other words, the human intellect, though born from very mundane preoccupations with tools and other practical means of survival, has developed into an instrument for approaching ultimate reality.

Science deals only with the intelligible facet of this reality. Are there not other facets, to be discovered by mental faculties other than intellect? And is it not conceivable that those faculties have, like intellect, grown from beginnings deeply rooted in biological survival, to become means of apprehending those other facets of ultimate reality? The question is worth raising. I shall return to it in the last chapter.

CULTURAL EVOLUTION

As we have seen in the preceding pages, the key to cultural evolution is *communication*.[5] An invention no longer dies with the inventor; it survives, to be copied and eventually modified by others, through models or instructions left by the inventor. Hunting and other food-gathering technologies, shelter building, artistic crafts, social mores, and countless other behavioral traits are similarly bequeathed for retention and possible improvement. The whole history of human civilization demonstrates the importance of transmission of cultural traits and of the cumulative process whereby past acquisitions are conserved, adapted to changing conditions, and enriched with new elaborations and revisions. This kind of evolution is sometimes called Lamarckian because it involves the inheritance of acquired characters. The image is wrong, however. Cultural traits are not inherited; they are conserved and transmitted, horizontally as well as vertically, by the spoken or written word, by objects such as works of art, and by other forms of communication. Virtually the entire human heritage has been built this way.

Several terms have been proposed to designate the "units" that are transmitted in cultural evolution. The most clearly defined such unit was introduced in 1976 by the British ethologist Richard Dawkins in his bestseller, *The Selfish Gene*,[6] under the name of *meme*, of which examples could be "tunes, ideas, catch-phrases, clothes fashions, ways of making pots or of building arches."[7] Memes, he suggested, have a life of their own. Once launched, they are replicated, though not always faithfully, by imitation or some other form of transmission and spread by jumping from brain to brain, in virus-like fashion. They are screened by a natural process of selection, the most successful ones—he mentions the "god meme" as example—advertising their success by the number of brains they inhabit or the number of times they appear in print or in recordings, in scientific jargon, their "citation index."

There is thus an analogy between memes and genes. Both are "selfish replicators," undergo mutations, and are subject to natural selection. Memetic evolution is, however, very much faster than genetic evolution, and it unfolds mostly horizontally, especially in the present world. In the past, songs may have been transmitted largely from mother to daughter. Today, they are spread worldwide by radio, television, tapes, and disks. Mothers learn them more often from their daughters than do daughters from their mothers.

The meme concept provides a valuable key to this process. Its originality is that it applies the Darwinian mechanism to the memes themselves. The success of a meme depends on its ability to induce its own replication in different brains. This property need not be associated with an evolutionary advantage. Many successful memes are neutral or, even, positively harmful. Yet, they spread. Think, for example, of drug-taking and other damaging behavioral traits.

The notion of meme has been adopted by a number of authors, including the American philosopher Daniel Dennett, who built his 1995 opus, *Darwin's Dangerous Idea,*[8] around it. In *The Meme Machine,*[9] published in 1999, the British psychologist Susan Blackmore has written the first "memetics textbook." In it, she analyzes the meme concept and its implications in detail and develops it far beyond its original definition by Dawkins. In her analysis, Blackmore sets great store on *imitation* as the principal means of meme replication and sees the ability to imitate as an almost exclusively human prerogative. In her words, "apes rarely ape,"[10] contrary to the view, held by many ethologists, that learning plays an important role in a number of animal societies. Blackmore also credits memes with a leading role, not only in cultural evolution, but even in biological evolution.

Memes, as already recognized by Dawkins and, with a different terminology, by many others, can influence genes, to the extent that the behaviors they command reflect on the survival and reproductive potential of the individuals who entertain them. An important contribution of the sociobiology school has been to relate cultural evolution to biological evolution. Useful cultural traits gave a selective edge to individuals genetically predisposed to exercise the trait most successfully, so that the genes responsible for the predisposition progressively spread in the population. Thus might a genetically determined skill at chipping stones become an evolutionary advantage once stone tools started to be used. How a process of this sort may have helped favor the expansion of the human brain has been touched upon earlier in this chapter.

Blackmore considers such an explanation inadequate. She advocates instead a process of "meme-gene coevolution,"[11] in which memes are in the driver's seat and, when they affect genes, do so by "changing the environment in which the genes were selected." She views this meme-driven process as more important than the classical sociobiological mechanism for such major evolutionary events as the expansion of the human brain and the development of language. In her opinion, the selective advantages of a bigger brain are insufficient to compensate for the higher energetic cost of building and running it and for the added risk to the mother at delivery. This argument ignores the very gradual pace of the processes involved. The energy spent in making and keeping in operation an additional ten milligrams of brain matter—the average increment per generation during the critical period of brain expansion—is surely negligible relative to the total energy expenditure of a primate, whereas the million-odd neurons the added bit of brain contains could very well, if properly situated, significantly modify some cerebral mechanism. Similarly, the very slow rate, in absolute terms, of the cranial enlargement accompanying brain expansion leaves plenty of time for the slow readjustments required by neoteny (see pp. 185–86).

Blackmore likewise dismisses Darwinian natural selection as accounting for the emergence of language. In discussing this problem, she rejects all sociobiological explanations in favor of a memetic one. "The human language faculty," she writes, "primarily provided a selective advantage to memes, not genes. The memes then changed the environment in which genes were selected and so forced them to provide better and better meme-spreading apparatus. In other words, the function of language is to spread memes."[12]

It does not behoove me, as an untutored onlooker, to pass judgment on differences that concern experts. Let me simply point out that if, as suggested above, the acquisition of language was a late event in human evolution, the long hominization process that preceded it could, as is generally accepted, have depended predominantly on Darwinian mechanisms, with the memetic mechanism slowly growing in importance as the brain became more complex and the meme pool richer. The emergence of language could have been a critical stage in this historical process, allowing memetic evolution eventually to outpace genetic evolution by several orders of magnitude. Thus could be explained (assuming it coincided with the use of language) the cultural explosion that was launched some 50,000 years ago. In today's world, memes clearly dominate, to the extent that they may even force genetic evolution to counter natural selection.

The Uniqueness of the Human Condition

As just reviewed, the advances of science have projected a new, deeply disturbing light on the human condition, which we always had every reason to consider unique. Compared to even the most advanced among the other animals, human beings exhibit a number of traits that distinguish them as *radically different*: rational thought, together with the ability to analyze, conceive, and understand abstract notions; the capacity to experience and express esthetic emotions; the feeling of moral responsibility; the power to love. The whole structure of civilization is built on these uniquely human qualities.

It now appears, with little room for doubt, that all the key properties that make us human are derived, through a continuous evolutionary process, from cruder traits present in a group of primates that existed on Earth some six million years ago and from which we are descended. The ancestral primates similarly derived their characteristic traits from less-endowed animals, as part of a progressive process that goes back to the earliest manifestations of psychic life. The lesson is humbling. It is also puzzling, in that it raises the question how a manifest discontinuity can arise through a continuous process. Evolution poses many questions of this kind, because it often yields only the finished products of a key transformation and eliminates intermediates. From fish to amphibian, from dinosaur to bird, the jump seems equally abrupt as from chimp to human. The scarcity of "missing links" long constituted one of the major objections against biological evolution and is still brandished as such in some creationist circles. The key variable here is *time*.

It is often said that the last step to human took only an instant of evolutionary time, the equivalent of the last two-and-a-half minutes if the history of life were reduced to one day. This is true. In human time, however, as already pointed out, the distance separating us from the last ancestor we share with other extant primates is huge: some 250,000 generations, as compared to fewer than 100 since the birth of Christ and fewer than 1,000 since Cro-Magnon man decorated the caves of Lascaux. Small step by small step, much can happen in such a long succession. If you are not convinced, visit a dog show. All the animals you see have originated by artificial selection in a mere ten thousand years. Even allowing for the more rapid reproductive rate of dogs, this still represents a much smaller number of generations than were involved in hominization.

Although the products of a continuous process, the discontinuities of evolution are quite real. Amphibians walk or jump; fish don't. Birds fly; reptiles don't. Humans do all sorts of things chimpanzees don't do, and can't. In this respect, in spite of the continuity through which it originated, the advent of humankind represents a true watershed.

For the first time in the long history of life on Earth, a species has emerged that is capable of uncovering and understanding the secrets of the universe and of using its acquired knowledge to consciously and deliberately manipulate the world, including other living beings and its own nature, for clearly defined purposes; a species endowed with unique intellectual, artistic, and social abilities; a species, in particular, invested with the redoubtable burden of moral responsibility. I shall come back to the significance of this event. But, before that, we must take a look at the organ where the most decisive changes responsible for the advent of humankind took place: the brain.

14. The Riddle of the Brain

BIOLOGICALLY, we are primates, close relatives of chimpanzees, with which we have more than 98 percent of our DNA in common. In terms of genes, the kinship is even closer, as part of the difference concerns "junk DNA," which has no coding function. Yet, mentally, the distance that separates us from our simian cousins is huge. In the past, this was explained by our having a *soul* that animals did not possess. Today, the explanation is that we have a *bigger brain*. What is it about this organ that explains the wonders of mental life?

In recent years, this question has become a central topic of research, sometimes referred to as the "last frontier," involving some of the best neurobiologists, psychologists, cognitive scientists, computer experts, and philosophers in the world. The field is actively fermenting, and I can give only an interested but unspecialized onlooker's view of its dynamic state.

From Brain to Mind

Mind is in the head, sustained by the brain. That much we know from everyday experience. What modern science has taught us in addition is that mind and brain are intimately connected, anatomically, functionally, and historically, by linkages that are beginning to be understood. The two are indissolubly linked, leading to the notion that thoughts, feelings, and all other manifestations of the mind are products of the activities of neurons in the brain. The concept is not new. The same was said two centuries ago.

"THE BRAIN SECRETES THOUGHT AS THE LIVER SECRETES BILE"
Thus declared the eighteenth-century French physician Pierre Jean Georges Cabanis. The German literature attributes a renal version of the saying to the nineteenth-century Dutch physiologist Jakob Moleschott, who is said to have written: "The brain secretes thought as the kidney secretes urine." In the climate of the times, these affirmations were meant as provocative attacks on the religious belief in an immortal soul. At present, the words have lost their incendiary character and their substance is accepted by most neurobiologists.

How could they be faulted? The proofs are there, indisputable, that no manifestation of consciousness is possible without the normal functioning of cerebral neurons. Let this functioning be impaired by lack of oxygen, or by a drug or trauma, and loss of consciousness inevitably follows. Many cases are known in which deterioration of certain specific mental aptitudes, speech, for example, or musical memory, can be related to strictly localized cerebral lesions. Starting in the nineteenth century with the observations of the French physician Pierre Paul Broca and of the German psychiatrist Karl Wernicke, whose names have been given to the speech centers, a detailed mapping of the brain has been established on the basis of such data.

Whereas there can be no consciousness without normal neuronal function, the inverse is far from being true. All the acts of our vegetative life, the coordination of our movements, and many other complex activities are managed by our nerve centers without our being aware of this control or even without our being able to affect it in any way. Many consciously initiated gestures progressively transform, through learning and practice, into unconscious automatisms. We laboriously learn to walk, to ride a bicycle, or to play the piano, finally arriving at a situation in which the control of consciousness has become more a hindrance than a necessity. Even in reasonings, it is often difficult to distinguish the respective parts of the conscious and the unconscious. Whereas the mechanisms of these phenomena remain poorly understood, their structural basis is known.

THE SEAT OF CONSCIOUSNESS IS THE CEREBRAL CORTEX
The Latin word *cortex* means bark. As applied to the brain, it designates a thin, specialized structure that covers the entire surface of the organ. The cerebral cortex consists of a sheet of grey matter (rich in cells), about three millimeters thick, characteristically composed of six superimposed

cell layers. A dense network of arborescences links the cells of these layers by a very large number of connections, which unfold transversely from one layer to another, and laterally within the same layer. This network is connected to the other parts of the brain and, thereby, to the whole organism by a tight mass of sensory and motor fibers (white matter). The former convey to it sensory impulses coming from all parts of the body, in particular the sense organs. The motor fibers send impulses to all the muscles. These inputs and outputs delineate on the cortex surface a set of specialized areas, which are now well mapped.

The cortex is the seat of consciousness; only signals that pass through it give rise to mental experiences. Below the cortex, numerous nerve centers bridge sensory and motor impulses by pathways that bypass the cortex and, for this reason, escape consciousness. The whole of vegetative life is thus regulated in an unconscious fashion. So are many automatic movements, such as those that command the position of the eyes, co-ordinate gestures, or control balance. Some of these automatisms are inborn. Others are created by learning, descending, so to speak, from the cortex, where they have been set into place, to deeper zones, where cortical surveillance dwindles.

It is remarkable that the structure of the cortex is essentially the same throughout the vertebrate series. What changes is its surface area, which reaches 2,200 cm^2 for the human cortex, forcing it to make numerous infoldings, or convolutions, in order to fit within the skull. The surface area of the cortex is about 500 cm^2 in chimpanzees. In rats, it is four to five cm^2, which, corrected for body weight, would amount to some 180 cm^2 for a human-sized rat. It goes on diminishing as we move down the animal scale, but there is the beginning of a cortex as soon as a true brain becomes distinguishable. Even in fish, there is a small cortical area, mainly linked with olfactory centers, which are particularly developed in these animals.

These facts suggest strongly that the characteristic, six-layer structure of the cerebral cortex is the generator of conscious experiences and that the richness of these experiences is somehow linked to the surface area of this brain structure. Expansion of the cortical area from 500 to 2,200 cm^2, for example, is what extended the range of problems soluble by the brain, from fishing termites with a twig to sending a man to the moon or engineering life.

According to neuroanatomists, the organization of the cerebral cortex is subdivided into functional blocks, or *modules*, associated laterally in the plane of the cortex. The richness of mental experiences thus appears to be linked to the number of such modules, somewhat like the performance of a computer is linked to the number of interconnected microchips of which it is composed.

Much detail has been added in recent years to our understanding of this functioning, especially in the visual area. A key factor made possible by the increase in the number of the brain's "microchips" seems to be the multiplication of parallel circuits carrying different aspects of the same information. Entry into the field of consciousness could be linked to some kind of resonance among those different circuits, by a phenomenon analogous, though of a different nature, of course, to what causes a crystal glass to burst into pieces under the influence of a sound whose frequency corresponds to one of its intrinsic frequencies, or a bridge to rupture when vibrating in phase with troops marching in step.

The underlying operations, in the course of which all kinds of circuits are tried and compared in more or less random fashion until the resonating combination emerges, are not conscious but could, when vigilance relaxes during sleep, "leak" into the cortical field. Dreams could thus, as first suspected by the founder of psychoanalysis, the famous Austrian neurologist Sigmund Freud, open a window into what is going on subconsciously in our brain.

Before moving on to the nature of consciousness, we must consider an important additional aspect of the functioning of the brain, namely its wiring. How does such a stupendously complex network of interneuronal connections arise in the course of development? Modern science has yielded valuable information on this topic.

WIRING OF THE BRAIN OCCURS EPIGENETICALLY

The brain is estimated to contain on the order of 100 billion neurons, each of which is linked to other neurons by an average of 10,000 connections. In total, this amounts to one million billion interneuronal connections. The diploid human genome contains some five to six billion base pairs. Thus, it is obvious that the wiring of the brain cannot be written into the genes. It must occur *epigenetically*, that is, by superimposed processes that take place during development. Genes only provide a general framework, which delineates the main features of the characteristic cerebral structure of the species. All the details of the interneuronal connections are created, within the limits of this framework, under the influence of inputs that reach the brain both from the body itself and from the outside world.

The mechanisms involved in this wiring have been analyzed by the French neurobiologist Jean-Pierre Changeaux and by the American Gerald Edelman, who, after completely elucidating the chemical structure of an immunoglobulin, has become actively involved in neurobiology. In brief, what happens, according to these two investigators, is that growing neurons continually establish transient connections with one another in essentially random fashion. The connections are rapidly undone, unless some outside influx causes them to be utilized, in which case they are stabilized. The neuronal network is thus established in the beginning by usage and continues to be so throughout life by learning. There are two important aspects to this mechanism.

First to be noted is its analogy with the fundamental mechanism of Darwinian selection. Chance creates a wide diversity of connections, among which are selected, under the influence of environmental factors, those that are retained and amplified. Hence the name of "neuronal Darwinism" given by Edelman to this mechanism.[1] Darwin is also cited by Changeaux in this sentence: "The Darwinism of the synapses takes over from the Darwinism of the genes."[2]

A second important aspect of this mechanism is that it highlights the capital role of communication in the psychic development of children. The manner in which we treat our children from the day of their birth onward, perhaps even earlier, literally shapes their brains and, thereby, their personalities. All prospective parents must know this lesson and draw the necessary conclusions. If they wish their children to develop a rich neuronal network, the condition of a rich personality, they must talk to them from the very first day, let them hear music, sing to them, cherish and caress them, attract their visual attention, give them toys of various shapes and colors; in summary, provide the children with a multitude of sensory stimulations thanks to which they will be able to build the innumerable neuronal circuits that underly the blossoming of mental life. One should not, as was often done, wait until a child develops understanding of language to start communicating with it. Even if it does not understand, it records. That is what is important.

Thus, much progress has been made and will no doubt be made in the future in our understanding of the brain on a purely phenomenological level. While important, these advances leave unexplained the actual nature of consciousness and that of its relations to neuronal functioning.

The Mysterious Nature of Consciousness

To most of us, consciousness is a self-evident fact. It is something that goes on in our brains and allows us to think, feel, make decisions, and speak and act accordingly. It is at the core of our deep-seated feeling of self as the person in charge of our destiny. It is the foundation of morality and of the underlying notion of personal responsibility. This intuition is now increasingly being challenged by a number of neurobiologists and philosophers, who see it as an illusion, a trick played on us by natural selection, an expression of something derisively called "folk psychology," as remote from the real thing as is "folk physics" from quantum mechanics. In the opinion of many of the greatest experts of our time, cerebral neurons do all the job and it makes no difference whether or not consciousness accompanies their activities. In the words of Crick, "you're nothing but a pack of neurons."[3] By this paraphrase of *Alice in Wonderland* ("you're nothing but a pack of cards"), the co-discoverer, with Watson, of the double-helical structure of DNA, who has converted to neurobiology, sums up the currently fashionable view on the subject. Neurons do it all on their own, whatever may go on in the mind at the same time. Implicit in this view is the notion that consciousness—if it exists at all—is no more than an *epiphenomenon*, some sort of emanation produced by, or associated with, the activities of certain parts of the cerebral cortex but totally unable to influence these activities.

IS CONSCIOUSNESS A SIMPLE ONE-WAY EPIPHENOMENON?
You may have the feeling of being in command of your ship and responsible for its running. But that, we are now told, is an illusion. When, at the end of sometimes agonizing ponderings, you finally resolve on saying "I believe," "I love," or "I will," or decide on a gesture or an action, what you call your "self" has nothing to do with it. It is no more than a spectator, a puppet mistakenly imbued with a sense of importance, perhaps even a mere phantom, while, in fact, your neurons pull all the strings. Extreme defenders of the theory even go so far as to maintain that consciousness is itself an illusion, without real existence.

An important implication of the epiphenomenon-illusion theory is that there is no free will. The defenders of this view realize that their affirmation undermines the very basis of moral responsibility, a cornerstone of human civilization. But they argue that whether we do, or do not, enjoy free will in reality makes no difference in practice. What count are the behaviors and wired-in connections built around the free-will illusion. Seen as an implacable consequence of the determinism of the brain-mind relationship, this notion is now accepted, albeit grudgingly, by a number of scientists and philosophers. It is even cheerfully hailed as "liberating" by some. In the words of Susan Blackmore, already cited in the preceding chapter, "there is no truth in the idea of an inner self inside my body that controls the body and is conscious. Since this is false, so is the idea of my conscious self having free will."[4] And she concludes: "In this sense, we can be truly free."[5] A number of other experts have, in one form or another, expressed the same opinion, concluding that, such being the truth, we have no choice—incidentally, *who* is denied the choosing?—but to accept it and act accordingly.

Those who refuse to accept an affirmation that contradicts their most intimate inner experience are sent back to Darwin. They are told that natural selection has favored the awareness of self, the feelings of freedom and responsibility, because these traits determined a social behavior that was useful to the preservation of the species. Prehumans genetically disposed to feel responsible with respect to the group had, collectively, a greater chance to survive and produce progeny than those who lacked this feeling. To be useful, the feeling need not correspond to reality. An illusion would do just as well.

True enough. But there can be no illusion without a victim. If being a victim of an illusion is itself an illusion, one arrives at an infinite regression of illusions, of the "Russian dolls" kind. Note that the defenders of the illusion theory use the same image to attack the notion of self, which, they maintain, supposes a self that looks at it, and so on, *ad infinitum*. The self, however, can look at itself, closing the series; whereas an illusion without a victim is inconceivable.

It therefore seems reasonable to start with the assumption—most would call it conviction—that consciousness is a real thing. We are conscious beings. The whole richness of our inner life is linked to this faculty. On it depends our ability to reason, to create, to enjoy, to suffer, to love, as well as the power we have to act in accordance with those perceptions. However illusory such abilities and powers may be, our being conscious of them is not an illusion. The heart of the matter is not whether consciousness exists, but what it does. Put more clearly, what we want to know is whether consciousness is purely *passive* or whether it can play an *active* role.

In general, neurobiologists, together with a few "materialist" philosophers, tend to support the former view on the basis of what they know of the functioning of the brain. What they see is, indeed, nothing but a "pack of neurons," interacting by a variety of physical and chemical signals and accomplishing stupendous feats of integration and coordination, most often without our even being aware of what they are doing. Take the cardiovascular system. At every instant, the brain and its associated nerve centers are deluged with messages coming from all over the body that may mean things like "send more blood," "oxygen crisis," "pressure reaches danger point," and the like. The messages are decoded and sorted, and appropriate responses are sent telling the heart to beat faster or slower, or the blood vessels to dilate or contract. All this complex interplay is never put into words or disclosed to our consciousness. It all takes place automatically, thanks to wired-in circuits, somewhat, though in enormously more complex fashion, like a thermostat adjusting the heat output of a furnace in relation to the temperature of a house. Many other physiological regulations and manifestations rely similarly on occult polyneuronal activities, which are beginning to be unravelled thanks to the powerful new techniques of neurobiology.

Even when consciousness is involved, the brain may carry out a considerable amount of hidden processing before sending the information to consciousness. In vision, for example, more than 30 separate channels record different aspects, such as outlines, movements, colors, shades, and so on, of the image projected on the retina. All these messages are integrated into a coherent picture by a process known as *binding*, accomplished entirely unconsciously. This process, which has been compared earlier to the physical phenomenon of resonance, seems to be related to a synchronization of the various impulses, leading to a characteristic 35 to 75 hertz oscillation, which Crick and his German-American coworker Christof Koch believe to be the basis of consciousness. The final result of this complex process is what we "see."

Even our reasonings may depend to a large extent on unconscious operations by which the neuronal phenomena that underlie concepts are sorted, combined, compared, weighed, and otherwise processed, before the outcome is delivered to consciousness to form what we experience as the substance of our thoughts. Add to this impressive sum of facts, which is growing almost daily, the notion that consciousness probably started with feelings associated with certain physiological activities, and you readily understand why consciousness is viewed by so many experts as an essentially passive epiphenomenon or correlate of certain polyneuronal processes, a manifestation that merely reflects the occurrence of those processes but lacks the power of influencing them in the manner implied by the existence of free will.

It has been objected that a purely passive consciousness can have no useful function and, therefore, no selective value. But this argument is not convincing. Suppose consciousness were just an inseparable concomitant of certain neuronal activities taking place in the cerebral cortex, a sort of "glow" or "hum," so to speak, that happens to obligatorily accompany those activities. Its selection as such would then need no explanation. The usefulness of the activities that generate consciousness suffices.

A more serious difficulty with consciousness is that it cannot be explained in known physical terms. It is an inner experience, inaccessible to investigation from the outside. Consciousness cannot be recorded. All that can be done is to record the underlying electrophysiological and biochemical manifestations, to note the accompanying behaviors, and to try to establish correlations between those manifestations and behaviors and consciousness phenomena. Even this requires the participation of the individuals who lend themselves to the experiments and inform us of what is going on in their minds. All we can know of the minds of others we know from their testimony and by analogy with our personal experience. This, of course, is the main reason why consciousness is ignored by all those who believe that only what can be "objectively" observed, analyzed, or measured is validly addressed by scientific research.

A beacon that can serve as a guide in the fog of theories and speculations is the fact that consciousness gives access to *abstract notions*, such as truth and falseness, beauty and ugliness, good and bad, comprehensibility and mystery, and that it allows us to communicate to others, by way of language and artistic creation, our personal vision of these notions. These facts form the main basis of *dualism*, a theory that attributes consciousness phenomena to an immaterial entity, soul, or spirit, bound to the material body but distinct from it.

DUALISM IS INTUITIVELY SATISFACTORY
BUT LOGICALLY INDEFENSIBLE

Instinctively, we are all dualists. We speak and act as persons, owners of a machine we call our body and over which we believe we have authority. Our entire affective and moral life is pervaded by this dualism, as when we complain that our body has betrayed us, regret having given in to its weaknesses, or, conversely, exult in our power to master them. But beware; this feeling, as we have just seen, may be only an illusion, a trick played on us by our neurons with the complicity of natural selection.

Many religions are dualist. Thus, Christian religions associate with the material and mortal body an immaterial soul that survives after death of the body. Here again, however, we must guard against simplistic ideas. Even though it is true that the immortal soul and the thinking self have long been considered as one and the same, and still are in the minds of many people, Christian thinkers are increasingly aware of the need to redefine the human soul in the light of discoveries in neurobiology.

Dualism is often linked to the name of the seventeenth-century French philosopher René Descartes, who defended a dualist conception of human nature in his celebrated *Discours de la Méthode*. It must, however, be noted that what is truly novel in the philosophy of Descartes is less dualism than *mechanism*. For Descartes, the human body, like that of all animals, may be considered a machine entirely subject to the laws of physics and explainable in terms of those laws. God has done no more than edicting the laws and creating the machine, subsequently letting nature take its course without other intervention. In this sense, Descartes rejects vitalism, which, without having been clearly formulated, was nevertheless implicitly present in all the thinkings of his time.

This conception was revolutionary. Descartes is well aware of this. As a good Catholic, anxious, moreover, not to suffer the fate of Galileo, who had just been condemned by the Holy Office, he insists on the purely imaginary character of his hypothesis. He presents it as a logical construction, entirely in accordance with reason, but hastens to add, in order to avoid getting into trouble, that things most likely happened differently.

In this perspective, Descartes's dualism is less a new theory than a minimal concession of his mechanistic philosophy to doctrine—to which, incidentally, he entirely subscribes. His dualism applies exclusively to human beings, who are the only possessors of a soul, the seat of reason, which distinguishes them from animals. He postulates an interaction between the soul and the body-machine, assumed to take place in the pineal gland, which, by its location in the middle of the brain, clearly betrays its central role. There, the soul communicates with what Descartes calls "animal spirits." Contrary to what this term might seem to suggest, these are material entities, present also in animals and described by Descartes as "a very subtle wind or, rather, a very pure and vivid flame, which, continually rising from the heart to the brain, will travel from there through the nerves to the muscles, and gives movement to all the limbs."[6] In animals, all is automatic. In humans, the soul manipulates the animal spirits in the course of their passage through the pineal gland.

Cartesian dualism has practically no defender left among contemporary scientists. There is a notable exception, the Australian John Eccles, an eminent neurobiologist, winner of the 1963 Nobel prize in medicine, who remained until his death in 1997 an unconditional exponent of a dualist theory of brain function. Needless to say, Eccles no longer spoke in terms of animal spirits manipulated in the pineal gland. He imagined the presence, at the level of synapses (the junctions between neurons), of "microsites" where neurons and consciousness could interact thanks to quantic fluctuations requiring no energy.

Phenomenologically, the mechanism conceived by Eccles has been met with skepticism by most cell biologists. As to the implication of quantum mechanics, we shall see that many investigators have tried to call on this discipline to explain the relationship between neuronal function and consciousness, without necessarily defending a dualist theory, at least explicitly.

Concerning this theory, I wish to point to what seems an intrinsic, logical flaw of Cartesian dualism. If matter and spirit are entities of truly different nature, the question arises as to how they can interact. This necessarily requires the participation of a hybrid form—a *"matter-spirit connector"*—sharing properties of both entities. But, if such is the case, matter and spirit join in a *single entity*. There can be no true dichotomy between the two. Thus, "hardline" dualism does not withstand logical analysis. Upon reflection, what lies behind dualism is a preconceived definition of matter. Even today, although dualism is widely rejected, the influence of this doctrine subsists in the conception most people have of matter.

WE MUST REDEFINE MATTER

According to a definition that goes back to Aristotle and represents the keystone of Cartesian dualism, we call matter what is inert, solid, and blindly subject to physical forces. We give another name to anything that does not fall within this definition. For the ancients, this was the case for many natural occurrences, such as astronomical phenomena, tempests, lightning and thunder, or volcanic eruptions, which they attributed to the strifes and caprices of imaginary gods, who combined magic powers with very human preoccupations. With the advances of physics, all these phenomena have been incorporated into the realm of matter.

Life was long viewed as an exception, but this is no longer so. Modern science has refuted vitalism by demonstrating that the basic phenomena of life are entirely explainable in purely physical and chemical terms. Thus life, in turn, has come to be seen as a manifestation of matter. This fact has not been recognized without difficulty and remains hard to accept by the average person, as it requires a widening of the *a priori* vision we have of matter as something gross, inert, and brute, and of life as "animated" matter.

There remains mind, the last bastion of animism. We have seen that this bastion is increasingly assailed by the advances of neurobiology. On the other hand, I have mentioned the logical problem raised by dualism. There is thus growing support in favor of a *monistic* (from the Greek *monos*, single) conception that tends to drive mind as well into the domain of matter. Even more upsetting than the rejection of vitalism, this conception forces us to widen our vision even further. We must loosen the stranglehold that the dualist intuition, reinforced by natural selection, has tightened around our definition of matter. Modern science has shown the way.

Before the invention of scientific instruments, we knew of matter only what is perceived by our sense organs, the products of a long evolution in the course of which natural selection has retained only what was immediately useful to the ability to survive and reproduce. We now know that the image of the world offered by our sense organs, if perfectly effective in everyday life, has little to do with true reality.

What appears to us as solid and impenetrable is mostly vacuum, within which atomic nuclei, which contain more than 99.9 percent of the mass of any object, are more distant from each other, relatively speaking, than are the planets of the solar system.[7] The illusion of solidity arises from the fields of force generated by the tenuous clouds of electrons that lie between the nuclei. The latter are nothing but enormously concentrated energy packets—according to Einstein's celebrated equation, $e = mc^2$—whose power, dramatically revealed in the infernos of Hiroshima and

, manifests itself more pacifically today in our nuclear power

Our other perceptions of the outer world are equally misleading. We saw in Chapter 3 that our eyes perceive only an infinitesimal fraction of the radiation emitted by matter. Our vision of the world would be totally different if another window were opened to us in the spectrum of electromagnetic radiations. A similar selectivity affects the sounds detected by our ears and the chemical substances discerned by our sense of smell.

Thus, our intuitive definition of matter is completely distorted by the filters our sense organs interpose between an object and us. It is an essentially pragmatic definition, based on the kind of information that proved most useful in the search for food, the fight against predators, and reproductive success. As means of knowledge, this information is almost valueless. It has needed the long road of scientific investigation to bring us progressively closer to true reality. This reality has proved so strange as to preclude any sort of mental representation, except by way of an abstruse, multidimensional mathematical formalism that is inaccessible to most of us and untranslatable in terms of ordinary experience, even by those who understand it.

These discoveries give a completely different meaning to the philosophical doctrine known as "materialism." Opposed to spiritualism, this doctrine has long been brandished, whether aggressively by its proponents or defensively by its opponents, as signifying that there is "nothing but" matter. Monism affirms the same thing, but *redefines matter as including the "spirit."* It is interesting to find this opinion defended, under the name of "metarealism," by a distinguished Catholic philosopher, France's Jean Guitton, who died in 1999. In a 1991 book titled *Dieu et la Science* (God and Science), which relates his conversations with two physicists, the brothers Igor and Grichka Bogdanov, Guitton writes: "If spirit and matter have a common origin, it becomes clear that their duality is an illusion, due to the fact that one considers only the mechanical aspects of matter and the intangible quality of spirit."[8] And he concludes: "Spirit and matter form a single and same reality."[9] Under the pen of a practicing Catholic considered one of the most eminent Christian thinkers of our time, such a sentence is both surprising and revealing.

THE QUANTUM MIRAGE

A number of experts, dissatisfied with the apparently inexorable determinism of the brain-mind linkage, have taken refuge in the indeterministic haze opened by quantum mechanics and by the famous "uncertainty relations" uncovered by the German physicist Werner Heisenberg, which, they believe, open the door to conscious choice without violating the laws of physics. To Kenneth Miller, the Catholic biochemist quoted in Chapter 12,[10] quantum indeterminacy forms the bridge between science and religion. It is the element of uncertainty deliberately introduced by God into the laws of His creation so that human beings can be free to choose between good and evil and bear responsibility for their actions.

Attempts have even been made to identify the cellular sites and mechanisms of these quantum events. I have already mentioned the "synaptic microsite" theory put forward by Eccles. The British mathematician and theoretical physicist Roger Penrose has, in collaboration with the American cytologist Stuart Hameroff, proposed a model based on "quantum gravity" and involving microtubules, which are tubular structures, about 25-millionths of a millimeter in diameter, made of protein units known as tubulins.[11] Details of this theory are too complex to be examined here. Let it simply be recalled that microtubules are present in all eukaryotic cells, not just in neurons. Thus, a unicellular protist or a single sperm cell would already possess the structural basis of consciousness. So would plant cells.

I must leave it to the specialists to judge the possible relevance of quantum mechanics to brain function. As a mere spectator, interested but lacking the required expertise, I can only make two remarks of a general nature. First, it seems to me that all explanations based on quantum mechanics retain a strong aroma of *dualism*. It is always the spirit that manipulates matter, with only the added clarification that it does so by taking advantage of matter's indeterminacy at the fundamental level. We are no longer dealing with the soul influencing animal spirits, but it amounts to the same thing.

The second point that strikes me is that quantum events are invoked mainly to "cheat thermodynamics." They are introduced as a means of breaking the deterministic chain of polyneuronal activities without the expenditure of energy, thus opening the door to free will. But free will does not necessarily mean "free lunch." There is no objective reason for assuming that the brain would generate consciousness and the ability to choose without spending energy. The opposite appears much more probable.

DOES THE MIND WIELD POWER?

Expressions such as "mental power" or "willpower" are part of everyday language. Yet, they have become almost dirty words in science, largely, I suspect, because they too readily raise the specters of telepathy, extrasensory perception, spoon bending, and other spurious claims. Not that such manifestations should be ruled out *a priori*. That would be unscientific. What distinguishes the paranormal from the normal is not its intrinsic impossibility—how often has a phenomenon been declared impossible until it was observed!—but the total lack of credible proof of its existence. Whenever a so-called paranormal phenomenon has been subjected to rigorous scientific control, either nothing has been found or a fraud has been uncovered. Adding the fact that the paranormal serves as a screen for a gigantic commercial exploitation of public gullibility, which sometimes, notably in the medical domain, may have dramatic consequences, one understands the hostility of scientists. In the case under discussion, however, we are not dealing with paranormal phenomena, but with perfectly normal phenomena in search of an explanation. This is very different.

In my opinion, to declare consciousness an irrelevant concomitant of polyneuronal functioning in the cerebral cortex, let alone dismiss it altogether from our preoccupations, for the sole reason that it is inaccessible to objective analysis, is not a satisfactory intellectual attitude. It ignores the whole gamut of subjective states; the countless sensations, feelings, moods, and emotions that color our lives; the thoughts, reasonings, speculations, and fantasies that flit through our minds; the ability we possess to focus our attention, to concentrate on a single topic to the exclusion of all others. It also disregards all the products and achievements of our mental activities; the innumerable expressions originating, through science, philosophy, poetry, music, the visual arts, religion, and other forms of culture, from our inner experiences; the increasingly effective tools and technologies, up to sophisticated computers, interplanetary spacecraft, nuclear power stations, biotechnologies, and other fruits of human ingenuity, developed with the help of our knowledge; all the richly textured social fabrics, which now envelop the entire planet, created by the sharing of experiences with the help of language and other means of communication. Finally, it belies our deep-seated conviction of being in charge of our fate and morally responsible for our actions. Such sacrifices for the sake of intellectual coherence and materialistic comfort seem to me excessive and unwarranted.

Looking at the question with the simple, if not simplistic, eyes of a biochemist, I see it as a *transduction* problem. On one hand, there are the neurons accomplishing their complex functions by mechanisms that are explainable in known physical and chemical terms and open to *objective* scientific investigation and description. On the other, there are the associated *subjective* manifestations of consciousness, of unknown nature but presumably likewise explainable, which can be uncovered only by introspection and interpersonal communication. Joining neuronal activities and conscious experiences, there is a special six-layered arrangement of the neurons involved—characteristic of the cerebral cortex—that seems to be the specific transducer of the ones into the others. Interconnected neurons do all kinds of things in our brains. But, as far as is known to me, only those of the cerebral cortex generate conscious experiences, presumably by virtue of the special way in which they are associated into modules, which, themselves, are linked into a dynamic web of highly specific topography.

Biology knows a number of transducers. To begin with, there are those that use metabolic energy to assemble ATP. Then, there all the others that, most often by way of ATP but sometimes also by other means,[12] convert this energy into other forms of chemical energy, as in biosyntheses; or into the active transport of molecules or ions, as in membrane transporters and pumps; or into mechanical work, as in myofibrils, cilia, and flagella,[13] or into light, as in the emitters of fireflies; or into electric discharges, as in the electric organs of torpedo fish. In all cases, *one form of energy is converted into another*. It seems not unreasonable to assume that, as with other biological transducers, the operation of the cortical transducers also involves an energy conversion. In fact, the opposite would be surprising, as it would amount to the generation of something without the expenditure of energy, in violation of the laws of thermodynamics (supposedly circumvented, as we saw above, by quantum effects).

It thus becomes at least plausible that consciousness represents some sort of *physically energized state* and that the particular configuration of neurons in the cerebral cortex serves as generator and supporter of that state. If so, what becomes of the energy thus invested? Its fate could be to be simply dissipated as heat, in which case, consciousness would indeed be a mere epiphenomenon. But the possibility may also be envisaged that a greater or lesser part of the energy could be *returned*—the transducer being *reversible*—to the neuronal mechanisms from which it has emerged. This would be the phenomenon of *mental causation* so vigorously denied by much of modern neurophilosophy.

Such notions fit naturally within biochemical thinking and should not, it seems to me, encounter serious objections. But, if we wish to account for our mental life and have it be more than an illusion, we must go one step further. We must assume that the state of consciousness may, between energization and de-energization, undergo certain *internal rearrangements*—the mind at work?—with as a consequence *modifications of the neuronal mechanisms by consciousness*. There would thus occur a sort of to-and-fro interchange between neurons and consciousness, in the course of which our thoughts and feelings would be elaborated with the help of the metabolic energy of the neurons.

The bothersome aspect of this kind of model is that it implies the existence of some unknown physical manifestation. But is that so reprehensible? The truth is that we do not know the nature of consciousness. To call it physical is no more than following the monistic principle that we have seen as logically necessary. To consider it energy-dependent is in accord with thermodynamic principles and more readily acceptable, at least to me, than the image of an entity that somehow escapes thermodynamic constraints by the magic of quantum fluctuations. The history of physics is replete with instances where advances in knowledge have imposed radically new notions. Just think of gravitation, atomism, electromagnetism, wave-corpuscle duality, relativity, matter-energy equivalence, elementary particles and their binding forces, not counting such exotic entities as quarks, muons, bosons, charm, and the like. Perhaps biologists, who were so successful in proving the nonexistence of "something else," should also, like the physicists, have the audacity to envisage something that is not included in the known properties of matter. Such audacity seems to me more valid, in any case, than relegating that something to the realm of the illusory simply because we do not understand it.

We may never know the key to the riddle of the brain. On the other hand, if, as seems to me almost mandatory, mind depends on some kind of energy-dependent physical manifestation, I see no reason why it could not some day prove accessible to physical interferences that would allow probing of the underlying mechanism. In the meantime, I prefer to entertain a hypothesis that accounts, without being implausible, for my most intimate experiences—including my feeling of personal freedom and responsibility—rather than trying to convince myself that those experiences, including the effort of convincing myself, are no more than an illusion created—how?—by a "pack of neurons."

LUCY'S REVERIE

Remember Lucy? She is the young female australopithecene whose fossilized remains were discovered in 1974, in Ethiopia, by the American Donald Johanson and the Frenchman Yves Coppens.[14] Now, go back in mind some three million years and imagine the living Lucy resting, after a long day's trek through the savannah. The hunt has been successful and a few fruits have added a welcome dessert to her meal. A dozen of her congeners of both sexes are lying around her, providing a warm feeling of togetherness and security. Contentment fills her body. Fleeting images of the day's activities pass through her rudimentary mind, concerned mainly with food, sex, care of the young, and companionship. Her eyes, through slowly closing eyelids, are absently gazing at the stars lighting up in the darkening sky.

Suddenly, just as she is falling into a slumber, she has a vision. Of sparks springing from two stones hitting each other, of shadowy forms on the walls of a cave dancing, in some flickering light, at the sound of grunts that seem to mean something, of larger-than-life figures raising sticks to the skies and wailing around a dead body, of a hairless newborn staring at her in touching helplessness. Awe, fright, and bewilderment alternate in the unformed recesses of her consciousness, giving place, in a brief moment, to a strange, evanescent sentiment of pure joy, never experienced before, inexpressible in the crude language with which she communicates with the other members of the band. Then, everything dissolves into sleep. The next morning, nothing but a vague memory of something uncommon, desirable, but unattainable, remains of her dream. She is back in her familiar world.

Today, we can interpret Lucy's reverie. With some 700 cm^2 of cerebral cortex surface area, she has been granted a nebulous glimpse of what would some day be accomplished with three times as much. The glimpse was crude, perforce limited to what she could attain by stretching her mental powers to the utmost. She could not possibly foresee our sciences, our arts, our societies, or our religions, share our thoughts or our emotions. She would have needed more cortical modules for that.

What, we may now ask, would happen if the surface area of our cerebral cortex should triple once more? If we extrapolate from the past, it is obvious that we can no more answer this question with our present mental means than Lucy could have done with hers three million years ago. We can only dream, like her, but without the vague premonitory glimpses our hindsight has given to our reconstruction of her reverie. All we can imagine is that those of our faculties that were born most recently have the greatest chance of developing further. We may assume, in this perspective, that science would become less laborious and impenetrable than it is today. Children would learn relativity theory in kindergarten, if learning was at all needed. Everybody would juggle black holes and superstrings as we now play with crossword puzzles, and handle the genetic vocabulary as we now do the alphabet. Minds would see reality beyond the appearances accessible to the sense organs and, perhaps, communicate with each other without need of speech or writing. Emotions similarly would be experienced without the necessary participation of sight or hearing. Perhaps, the blissful vision of love claimed by some rare mystics will be everyday experience.

All this, of course, is fanciful and necessarily shaped by elements of present reality and experience. By definition, our vision cannot include what to us is unthinkable. The difference between 6,000 and 2,000 cm^2 of cerebral cortex surface area is most likely to be as drastic as the difference between 2,000 and 700 cm^2. Whatever it implies, this thought is worth keeping in mind as we look into the future of our species and of the living world. This we shall do in Chapter 16. Before we do so, however, we must consider some recent technical developments that radically alter any sort of view we may have of our future.

15. Reshaping Life

W HEN, IN 1978, THE Nobel prize for medicine was awarded jointly to a Swiss geneticist, Werner Arber, and two American biochemists, Daniel Nathans and Hamilton Smith, this event was heralded by the following comment from Swedish television: "Their research opens up the possibility to copy human beings in the laboratory, to construct geniuses, to mass produce workers, or to create criminals."

This typical example of media hype has, less than 25 years later, come close to reality. What the slow and laborious process of evolution has been doing for nearly four billion years, by letting natural selection screen off, among myriad random mutations, the rare variants that happened to fit some local environmental conditions, can now be accomplished almost at will and almost in no time by a human-led engineering procedure that chooses and creates a particular genetic combination to fit a set of predetermined conditions. There is still a big element of uncertainty in this newly gained mastery over life because of technological hazards and of limitations in our predictive abilities. But the tools are available, wrestled from nature by human ingenuity, and now mass-produced for anyone with the required know-how to use.

THE MAGIC TOOLBOX

Like so many other recent technical developments of major importance, genetic engineering is the unforeseen fruit of basic research carried out for the sole purpose of solving a problem posed by nature. It all started in the early 1960s in Geneva, where the young Arber, freshly returned from a stay in several prestigious laboratories in the United States, became intrigued by some strange aspect of the interaction between bacteria and phages (short for bacteriophages, which are viruses that infect bacteria). The phenomenon involved, called *restriction-modification*, was odd. But rare were the investigators it kept awake at night. As it happened, it had tickled Arber's curiosity. In a few years' time, he had found the explanation. What it boils down to, leaving out technical details, is that certain bacteria defend themselves against phage infection by breaking down the invaders' DNA without damaging their own. Their trick is to attack the phage DNA by means of an enzyme that cuts the molecules at certain precise sites, identified by a short sequence of bases, and to protect their own DNA by chemically modifying those same sites (by the addition of methyl groups) in a manner that renders them resistant to cutting by the enzyme.[1] Later, Nathans and Smith isolated and characterized the first cutting enzymes, known as *restriction enzymes*, and established their specificity.

The reason this work turned out to have such a revolutionary fallout is that restriction enzymes, of which several hundred distinct varieties are now known, each with a different specificity, make up an extraordinary set of genetic tools. Previously, the only "scissors" available for cutting DNA were "illiterate"; they split the chains at almost any site and fragmented them, as would a shredder, into innumerable different pieces with which nothing useful could be done. Restriction enzymes are scissors that "can read." Imagine, for example, attacking the preceding paragraph with a kind of scissors that cuts only the sequence "tion" between "ti" and "on," another that cuts "sites" between "si" and "tes," and so on. The first kind will cut the paragraph into eight pieces, always the same for all copies of the book. The second one will produce five pieces. If the specificities are more stringent, the number of pieces will be smaller. For example, if the first scissors recognize "ction" instead of "tion," five instead of eight pieces will be produced. Note further that if your scissors should recognize roman type, but not italics, you could protect the paragraph simply by italicizing the sensitive sites. This, figuratively speaking, is how modification works.

Armed with scissors of this sort, you would be able to identify any book in your library without reading it, simply by counting the different kinds of pieces and measuring their sizes. It is clear that no other book would give the same pattern. Another pair of scissors would give you another collection of pieces, also characteristic for each book. As it happens, handy techniques are available for separating and sizing DNA fragments in a single operation, making it child's play to count and measure the different fragments obtained by the action of a given restriction enzyme on a given sample of DNA. With this simple method, known as *restriction fragment length polymorphism* (RFLP) determination, the owner of any sample of DNA can be identified with near certainty, because there are enough differences in the DNA sequences of different individuals to make the RFLP pattern of each essentially unique.

Thus, a first fruit of the discovery of restriction enzymes has been a means of identifying individuals by their *genetic imprints*, which are at least as specific as fingerprints and are beginning to replace the latter in forensic science. Murderers and rapists can now be identified from traces of hair, blood, or sperm left at the scene of their crime. Soon, perhaps, if certain lawmakers have their way, every person's genetic imprint will be on file in a central bank as a permanent, unerring means of personal identification.

Restriction enzymes have also proved valuable for DNA sequencing, as small DNA stretches are much more easily sequenced than the long, natural molecules. If at least two restriction enzymes of different specificity are used, the entire sequence can be reconstructed from the sequences of the fragments, by taking advantage of overlaps between the sequences of fragments obtained with different enzymes.

Particularly important applications of restriction enzymes became possible with the help of two additional tools, namely a copying enzyme (*DNA polymerase*) that can be used to make any number of copies of a given DNA sequence, and a stitching enzyme (*ligase*) that allows the linking together of DNA stretches in any order one desires. Thanks to an ingenious method, known as the *polymerase chain reaction* (PCR), devised by the American Kary Mullis, the copying enzyme can even serve to selectively fish out and amplify DNA molecules present in extremely small amounts in a highly complex mixture.[2] All that is needed is knowledge of a small part of the sequence of the DNA to be amplified. Another neat device, due to the late Anglo-Canadian chemist Michael Smith, is the technique of *site-directed mutagenesis*, already alluded to in Chapter 7, which makes it possible to change the copied sequences at specific sites. As to the stitching enzyme, it can be used to construct made-to-order genes with selected pieces of natural, artificially modified, or synthetic DNA, to attach genes to chosen carriers, or to process them in any other way one wishes. Almost anything I can do on my word processor with parts of this book can be done today with genetic texts.

Certain viruses whose genome is made of RNA, instead of DNA, have enriched the toolbox with two RNA-processing enzymes, one that copies RNA molecules (*RNA replicase*) and another that reverse-transcribes them into the corresponding DNA molecules (*reverse transcriptase*).[3] It has thus become possible to start also with RNA molecules; these can be amplified at will or converted to DNA molecules carrying the same information, which can, in turn, serve in all the manipulations referred to earlier.

In addition, a number of means have been developed for introducing genetic material into cells in such a way that the cells handle the foreign genes like they do their own. These means range from simply exposing the cells to the foreign DNA, hoping for the best, to highly sophisticated insertion procedures, including micromanipulation and, especially, the exploitation of natural gene carriers, such as viruses and phages, as well as small transferrable circles of DNA, called *plasmids*, present in many bacteria.

Using the Tools

Genetic toolboxes are now produced industrially and used in hundreds of laboratories throughout the world. They have opened the way to the most powerful, and perhaps the most disquieting, technology ever devised, which is in the process of transforming the future of life on our planet. First applied to bacteria and other unicellular organisms, the new biotechnologies have been progressively extended to plants and then to animals. They are now beginning to invade the human domain.

MICROBES

The simplest application of biotechnologies, now supplanted by enzymatic copying, is *gene cloning*, that is, their amplification. Bacteria are chosen as recipients for the genes because they multiply rapidly. After genetic implantation, the bacteria are allowed to proliferate; in so doing, they multiply the inserted genes at the same time.

In a more complicated application, conditions are chosen so that the implanted genes are abundantly expressed into the corresponding proteins. These techniques are predominantly applied to bacteria, which most readily lend themselves to them. But unicellular eukaryotes, such as yeasts, are also sometimes used, when advantage is to be taken of certain special mechanisms, for example, the secretory machinery. Organisms programmed in this way become factories for the production of proteins encoded by the implanted foreign genes. Several human proteins, including insulin and growth hormone, are now made industrially in this way. In addition to their economic value, these techniques can serve to avoid grave accidents that can occur when human tissues are used as sources of materials. Some years ago, in France, a number of children treated with growth hormone isolated from human pituitaries (a gland present in the brain) died of Creutzfeldt-Jakob disease, the human form of mad cow disease (see Chapter 2). It so happened that some of the pituitaries used for extraction of the hormone came from victims of the disease. In everyone's memory are the tragedies caused in France and other countries by transfusion of AIDS-contaminated blood to hemophilia patients. These heartbreaking events would have been avoided if the coagulation factor needed by the patients had been produced by genetically engineered bacteria.

PLANTS

For the genetic modification of plants, nature provided the means in the form of a bacterium, *Agrobacterium tumefaciens*, a name that, literally, means "field microbe that causes tumors." This is exactly what this microbe does. It induces *crown gall*, also called plant cancer, the kind of unsightly excrescence that can be seen to grow on many trees and other plant species. The microbe causes tumors by introducing into plant cells the copy of a DNA sequence present on a large plasmid, or piece of extrachromosomal DNA. This DNA sequence bears the cancer-producing genes and becomes integrated into the plant genome so that the genes can manifest their activity. Thanks largely to the efforts of two Belgian investigators, Jeff Schell and Marc van Montagu, the mechanisms underlying this remarkable phenomenon have been unravelled and turned into a powerful technology. The plasmid involved has been deprived of key cancer-producing genes and engineered to carry foreign genes, which can thus readily be introduced into plant genomes, with the bacterium serving as injection device. This technology has been greatly aided by the fact that, unlike animal cells, many differentiated plant cells can easily be caused to revert to an embryonic stage from which whole new plants can grow. The products of such manipulations are *genetically modified organisms* (*GMO*), also called *transgenic*.

One of the first transgenic plants created in this way was a variety of tobacco that secretes its own insecticide, thanks to implantation of a gene, called Bt, extracted from a bacterium, *Bacillus thuringiensis*, that naturally produces a potent insecticidal toxin. Several other plant species have been similarly engineered to release this bacterial toxin into their immediate environment, thereby protecting themselves against noxious insects. Another dramatic first was a species of maize, followed by several other crop plants, engineered to be resistant to a given commercial herbicide. This trait allows use of the herbicide under more acceptable environmental conditions and without harm to the modified plant.

There is almost no limit to what the new technology can accomplish. Among the genetically modified plants that have been created, are being developed, or are contemplated, in addition to those already mentioned, are species that have been rendered immune to a number of pathogenic viruses; plants capable of using atmospheric nitrogen and thus no longer in need of nitrate-containing fertilizers (which are by themselves a major source of environmental pollution and, through their industrial production, significant consumers of energy and contributors to global warming by released carbon dioxide); plants with stronger resistance against climatic rigors, such as cold or drought; male-sterile species forced to reproduce by hybridization, which gives rise to more vigorous offspring; yet others with enhanced natural properties that ensure higher crop yields, greater nutritional value, improved conservation after harvesting, or more abundant production of substances that can be used for the making of biodegradable plastics, lubricants, detergents, packaging materials, drugs, and other useful products. Thus, a tomato with delayed softening upon storage has made headlines. So has "golden rice," greatly enriched in vitamin A and iron, expected to help fight malnutrition in many underprivileged parts of the world. In addition, plants have, like microorganisms, been converted into factories for the industrial production of a variety of substances, including vaccines as well as animal and human enzymes, hormones, antibodies, and other proteins. It is no exaggeration to state that genetic engineering is in the process of revolutionizing agriculture and many manufacturing industries.

Developments of this sort can't be accomplished without huge investments in research. To ensure adequate returns on their expenditures, manufacturers of improved seeds have devised an ingenious but much-criticized method for protecting their products. It involves implanting a so-called terminator gene that renders seeds sterile. Farmers who use the products can no longer follow the age-old method of collecting their own seeds after each crop and sowing them. They must buy them anew. Although hailed also as a means of preventing the undesirable spread of modified genes, the terminator technology has now, under pressure of public criticism, been abandoned by its promoters.

ANIMALS

Animals, especially mammals, offer greater challenges to genetic engineering, because their differentiated cells do not readily revert to an embryonic stage and their egg cells, deeply implanted in the womb, are inaccessible to manipulation. The answer to the problem came from research on *in vitro fertilization* (IVF). In this procedure, an unfertilized egg cell is removed from a female's ovary and exposed to sperm in a laboratory dish. The fertilized egg is then implanted in the female's uterus, where it goes through a normal process of development. Sometimes, the egg cell is allowed to undergo a few divisions *in vitro*, following which the resulting early embryo or cells removed from it (called *totipotent* because they are capable of differentiating into every type of cell present in the body) serve for implantation. The intermediate, "dish" stage lends itself to a variety of manipulations, including gene transfer. It is also possible to produce so-called "knock-out" organisms, in which certain genes are inactivated.

Such interventions have proved invaluable in basic biological research, where genetically modified organisms are now used on a large scale in the investigation of a wide range of problems. The new technologies also have important practical applications. For example, transgenic animals are beginning to replace microorganisms as protein factories. In an experimental farm in Virginia, a sow called Genie has for several years produced daily in her milk, in amounts equivalent to the content of several thousand liters of human blood, a coagulation factor, named protein C, that is missing in a number of patients. By the time this book appears, a whole gamut of human proteins will be produced by appropriately programmed relatives of Genie or members of other similarly engineered species.

The crowning—provisionally?—achievement has been the *cloning* of animals, that is, the production of animals almost identical genetically to a chosen individual. In this technique, the nucleus of an egg cell is removed and replaced by the nucleus of a somatic cell[4]—from the intestine, for instance, or the mammary gland—of the organism to be cloned. The egg cell produced in this way contains two sets of chromosomes, just like a normally fertilized egg cell. But, contrary to the normal case, the chromosomes do not originate from two parental germ cells, of which each has provided one of the two sets, but from the differentiated cell whose nucleus has been used. Thus, the individual born from the development of the renucleated egg cell in a surrogate mother is a *genetic copy of the donor of the nucleus*, except for mitochondrial genes (see Chapter 10) and other forms of cytoplasmic heredity (see Chapter 2), which are provided by the recipient egg cell.

This technique was first successfully applied to amphibians, in the 1970s, by the British biologist John Gurdon. Then a Swiss investigator claimed to have cloned mice, but later had to retreat in some confusion, perhaps unjustly in view of what is now known of the hazards of the technique, when his results proved not to be reproducible. Cloning made a spectacular comeback in 1997, with the announcement, by a Scottish laboratory directed by Ian Wilmut, of the birth of the now world-famous Dolly, a ewe six years younger than its almost identical twin, the donor of the nucleus used to replace the egg's own nucleus. Since then, mammalian cloning has produced mice, calves, goats, and pigs. Application of the technique to primates has proved more difficult but will presumably be successfully accomplished some time in the future. Human cloning is already being contemplated, in spite of strong ethical objections (see below).

Note that the technology still remains highly problematic. Even under the best conditions, many attempts at cloning are unsuccessful or yield severely abnormal offspring. Dolly, for example, was the single product of 277 attempts. In general, the present success rate varies between 0.1 and 1.0 percent. These difficulties are understandable. In reality, what is puzzling is the successes, rather than the failures. As was seen in Chapter 11, body cells originate from embryonic cells by progressive differentiation, a process in which only certain genes—such as those that specify a brain, liver, or skin cell, depending on the cells' location in the developing embryo—come to be expressed, while all others are silenced. Thus, when the nucleus of such a cell is used for transfer, it must be *deprogrammed* so as to recover all the potentialities needed for the formation of the cells of the cloned organism. This reversal is beset by many difficulties.[5] The fact that cloning can be done shows that complete deprogramming of a body cell nucleus is possible, presumably thanks to the special environment provided by the egg-cell cytoplasm. However, the stage at which it can occur may be highly critical, thus accounting for the many failures.

Another problem with cloning, first disclosed by Wilmut himself, Dolly's creator, is that the implanted nucleus may carry signs of the age of the donor cell in the tail parts of its chromosomes. Called *telomeres*, these tails consist of a large number of TTAGGG repeats attached to one of the DNA strands at each chromosome end. It so happens that telomeres are incompletely reproduced when DNA is replicated before each cell division, so that they become progressively shorter generation after generation, until their small size renders further division impossible. This is a major mechanism of *cellular senescence*. Germ cells possess an enzyme, called *telomerase*, that specifically repairs telomeres. This enzyme is absent in normal body cells but reappears in cancer cells, thus accounting for the "immortality" of these cells. Since telomerase presumably was present in the cytoplasm of the recipient egg cell, one could have expected telomere repair to be part of the rejuvenation process undergone by the implanted nucleus. Contrary to this expectation, a May 1999 report by the Wilmut group disclosed that Dolly's telomeres were about 20 percent shorter than those of a normal lamb of the same age and had the length expected for the implanted udder cell nucleus. The ewe thus seemed fated to age abnormally fast and eventually to catch up with her nuclear parent, dashing dreams of eternal youth through one's clones. This, apparently, has not happened. Dolly has shown no sign of premature aging and has gone through two normal pregnancies.

Although animal cloning remains a hazardous process, there is little doubt that, as research progresses and the underlying phenomena are better understood, increasingly effective means of controlling each stage will be discovered and cloning will become increasingly easy, safe, and reproducible. It has already led to major business ventures and to the inevitable conflicts about patents. Among the major benefits expected from the new technology are a much more rapid multiplication of exceptional stocks than by conventional breeding techniques and, especially, of transgenic animals engineered to produce useful substances in their milk or urine or to provide immunocompatible organs and tissues for human replacement therapy. The neologism *"pharming"* has been introduced to describe this new form of agribusiness.

HUMANS

In view of their importance, applications of the new biotechnologies to human beings will be examined in a special section at the end of this chapter.

NATURE ADVOCATES TAKE ISSUE

Predictably, the developments of biotechnology have created a great deal of commotion among environmentalists, politicians, economists, ethicists, and, through them, the general public. Interestingly, the first qualms were expressed by the scientists themselves.

ASILOMAR

As early as January 1973, a number of leaders in the burgeoning field of molecular biology met in Asilomar, California, convened by one of the pioneers of the new technologies, the American Paul Berg. The aim of this meeting was to examine the potential dangers of genetic manipulations, which had just been shown to be feasible on bacteria. As a historic first—the anxieties of physicists about nuclear energy came mostly after the bomb—the participants decided on a one-year moratorium on any new experiment of the kind, to allow enough time for the in-depth analysis needed to reach an informed decision.

No one at the time of "Asilomar" had any idea of the whole gamut of interventions that would eventually prove feasible. The main worry concerned the accidental creation of dangerous, pathogenic bacteria or of carcinogenic viruses. After one year of reflection, covered by the moratorium, it was concluded that the risks of such accidents were negligible and that they could be avoided in any case by doing the experiments in the kind of contained laboratories in which highly virulent organisms, such as the Ebola virus, are handled. In spite of drastic protections of this sort, application of the new techniques met with strong opposition—largely ignited, ironically, by the scientists' commendable caution—even in certain scientific circles. Only by a hair's breadth was a decision by the mayor of Cambridge, Massachusetts, avoided; he wanted to ban the use of genetic engineering techniques within the entire territory of his borough, where are situated such prestigious institutions as Harvard University and the Massachusetts Institute of Technology (MIT).

THREATS TO THE ENVIRONMENT?

As of today, the first fears have been shown to be unfounded. The new technologies are universally authorized and put into practice in academic and industrial laboratories all over the world. But this does not mean that their products are accepted without objection. On the contrary, they are rejected all the more strongly because they consist not only of genetically modified microbes, but also of plants, animals, and—who knows?—soon human subjects. In a large number of countries environment-defense groups and the related "green" parties fight fiercely, some even violently, against biotechnologies.

The opposition has become particularly virulent—and successful—in Europe, to the point of creating a major, emotionally laden conflict, with considerable ideological, political, and economic overtones, pitting the main European countries against the more tolerant United States. Highly publicized by the media, the issue has become a topic for endless discussions, reports, commentaries, and other exchanges, which, more often than not, resemble what the French call a "dialogue of the deaf."

Some opponents of biotechnologies base their attitudes on an almost religious respect for nature, which, they hold, should not be tampered with on any account. In the words of the heir to the British throne, Prince Charles, who has not hesitated to join the fray, biotechnologies are taking humankind into realms that "belong to God and God alone." Scientists should not try to splice together genes that "God deliberately kept apart." Many others, without going so far as to invoke divine will, share the same feeling.

Leaving aside the emotional or irrational elements of the conflict, let us consider the scientific arguments put forward most often by the adversaries of the new biotechnologies. Their main concerns deal with possible ill effects from the introduction of transgenic organisms into the natural milieu. Some worry, without citing any specific effect, about upsetting fragile ecological equilibria refined by millions of years of evolution. This kind of objection ignores the fact that humans have not waited for genetic engineering to remodel nature and upset natural balances in a much more anarchic and blind manner than is contemplated by genetic engineers today. The varieties of maize, wheat, soya, and rape that cover our fields are only remotely related to their wild ancestors. The wolves and jackals that prowled around Cro-Magnon caves would be hard put to recognize a descendant in a chow-chow or a St. Bernard, just as prehistoric hunters would be astonished by the horses, cows, pigs, and other animals we breed on our farms.

But there is an important difference, we are told. Our ancestors used only *natural* means, such as hybridization and selection by the choice of progenitors. What is contemplated today is done with *foreign* genes, whose effects are difficult to predict. Take, for example, plants that produce their own insecticide thanks to implantation of the Bt gene. Is there not a risk that the insecticide be carried beyond the boundaries where its action is desired, there causing damage? This possibility materialized when, in May 1999, a report from Cornell University announced that pollen from maize engineered to produce the Bt toxin could, under laboratory conditions, kill Monarch caterpillars. Although the extent to which similar effects could occur in the wild is generally taken to be minimal, these results have had a considerable impact on the public reaction to GM crops. Indeed, the Monarch has an almost emblematic value in the United States. It is a brilliantly colored butterfly of large size, whose swarms cross the country in spectacular migrations between Canada and Mexico.

Another frequently formulated fear is that released pesticides might hasten the creation of resistant pests. According to all that is known, this is a very real risk. Several insect species resistant to the Bt toxin have already been created in the laboratory, and at least one has been detected in the wild. But this is a problem with all pesticides. Remember the resistance of mosquitoes to DDT mentioned in Chapter 12. The advantage of transgenic plants, according to their advocates, is that they allow replacement of the now-massive usage of pesticides by a more localized production, restricted to the immediate neighborhood of the plants. The risk of resistance remains but is more circumscribed.

The radical solution to the problem would, of course, be to totally ban the use of pesticides. This is what is recommended by the practitioners of what has quaintly become known—is there any other?—as "organic" farming.[6] The public infatuation with "organic" food products testifies to the success of this approach. But one may legitimately wonder what would happen to "organic" cultures if they should no longer be surrounded by the protective screen formed by the more conventional cultures and their pesticides. How long would it take, should the entire surface of the United States or of Europe become "organic," for the Colorado beetle, phylloxera, and other scourges of the past to surge up again?

Environmentalists also worry about the possibility that foreign genes might accidentally pass from engineered organisms to other species, either by horizontal transfer or by hybridization (crossing between species). For example, a variety rendered resistant to herbicides by transplantation of the appropriate gene could confer this resistance to some wild weeds, which would thus become refractory to our means of eradication. Such risks are not inconceivable and warrant attempts at prevention. Various isolation techniques are indeed applied to this effect. Note, however, that horizontal gene transfer has so far been observed only with bacteria. If it occurred frequently in the plant or animal world, this would have been known long ago. There remains, of course, the risk of transfer by hybridization.

The creators of transgenic species are far from insensitive to these arguments and surround themselves, for this reason, with multiple precautions in their open trials. In addition, no serious accident has been recorded so far where transgenic crops have been cultivated. Thus, in the United States, more than eight million hectares are covered with cultures of Bt toxin-producing transgenic maize without any ill effect having been observed, including on the Monarch butterfly. This has not prevented the adversaries of transgenic cultures from virulently pursuing their opposition, up to the point of physically destroying the cultures in certain countries, notably in France and Brazil, enjoying almost total impunity thanks to the support they have received from public opinion. As I write this book, the future of transgenic cultures is seriously threatened in Europe, and the movement seems to be gaining momentum in the United States.

Note that the criticisms and fears not only address transgenic plants. Animals are also targeted. Thus, a variety of salmon with accelerated development, created in Canada by implantation of a gene coding for growth hormone, has drawn the fire of nature defense groups; they fear that the release, accidental or deliberate, of the transgenic animals into the wild may cripple the species, should the transgenics be sexually more efficient and their progeny weaker, or, on the contrary, may lead to an exaggerated success of the species, which would then impoverish the

milieu by its voracity. These are mere conjectures, to be sure, but sufficient to inflame public imagination.

Opponents of GMOs do not just worry about environmental risks;
they see the quality of our food as threatened. As often in such cases,
the movement of opposition against transgenic foods has crystallized
around a single word.

"FRANKENFOOD"

This neologism, born in the United States, combines the word "food"
with the name of the evil Dr. Frankenstein, the creator of monsters in a
celebrated novel by Mary Godwin, the poet Shelley's second wife. The
term, which clearly says what it means, has become a rallying sign for
all those who oppose the sale and consumption of transgenic foods. This
opposition is explained in major part by hostility toward the methods
whereby these foods are obtained. But it rests also on the fear that the
new foods may cause toxic or allergic reactions. Such risks obviously exist,
and they have not awaited modern biotechnologies to manifest themselves. Many natural plants produce toxic or allergenic substances. It is
thus perfectly possible that certain transgenic plants may do the same.
In the meantime, no seriously validated case of such an accident has so
far been reported. On the other hand, the opinion climate is such that
transgenic food products are probably those for which the greatest precautions are taken before they are put on the market.

Nevertheless, the opposition to Frankenfood has become so strong
that most European countries would either ban GM foods outright or,
at the very least, require that any food product "tainted" by GM material
should be so labelled in order to allow informed choices by consumers.
The movement has reached the United States, where the Food and Drug
Administration (FDA) is increasingly urged to demand similar labelling.

The debate has long ceased to be strictly scientific and is now integrated within political conflicts, some of them violent, that involve the
notion of "globalization." The fact that advances in the transgenic domain are mostly a prerogative of powerful, multinational companies suffices to unite against them all those who, for one or another reason,
accuse these companies of capitalistic exploitation of the poor, especially
in developing countries. And yet, in the opinion of many impartial experts, those are the very countries, with their populations desperately

exposed to famine and malnutrition, their millions of children dying or wasting away each year for want of adequate food, that would gain most from production methods that produce higher yields or richer foods without the heavy and unhealthful use of pesticides. Remember golden rice. It would be deplorable if the peoples who need them most should be deprived of those benefits for ideological reasons.

WHAT ARE WE TO THINK OF ALL THIS?

In all objectivity, GMOs do not deserve the opprobrium they are stricken with. Unfortunately, the opposition mixes sound arguments with uncritical and sometimes erroneous, pseudo-scientific declarations, blending them into a mishmash that appeals to a poorly informed public readily seduced by a romantic, Rousseau-esque "return to nature" philosophy and haunted by the old myths of Prometheus, Pandora, and the apprentice-sorcerer. This opposition has consolidated, especially in Europe, into a powerful political movement with which it has become necessary to compromise.

Let it be clear, there is no rational support for the kind of mystique that animates the more emotional adversaries of these new technologies. "Mother" nature is neither wise nor benevolent; nor does she have any allegiance to the human species. Scorpions, poisonous toadstools, and AIDS viruses are as much objects of her solicitude as are butterflies, life-saving molds, and poets. Nature is governed entirely by natural selection according to an intricate network of influences that pit the conflicting interests of different organisms against each other (struggle for life), within the constraints imposed by their interdependence (ecosystems). Humans have emerged from this blind interplay, like all other living beings, but with the unique privilege—and burden—of a brain developed to the point that it can elucidate the mechanisms underlying this interplay and, thereby, devise means of manipulating them.

This revelation is entirely new. A mere 50 years ago, what is now done routinely with commercially available "kits" was not only beyond the means of biological science, it could not even be imagined, except by science-fiction writers unconcerned with feasibility. And here we are, suddenly faced with the presence in our midst of a small minority of wizards invested with exorbitant powers over the living world and liable to put their expertise at the service, not necessarily of the sole common good—not easily defined, anyway—but also of powerful political or financial interests. No wonder the rest of humanity is bewildered and frightened.

Innovation has always had this effect. Resistance to change is part of human nature, perhaps etched into it by natural selection. Mechanical looms, steam engines, railroads, and many other inventions that we now take for granted were all developed against strong opposition. For all we know, perhaps even the wheel was greeted with suspicion when it was first invented. In the present case, distrust is intensified by the fact that life itself, mysterious to most and sacred to many, seems to be under threat. Yet, there is no going back—and no reason for doing so. Biotechnologies are here to stay. The question is not whether we should respect a motherly nature that exists only as myth. What counts is that we no longer have to leave it to chance alone, including the vagaries of human conduct, to direct the course of evolution; we can now, knowingly and responsibly, take a hand in this process. The problem is not *whether* we should do something, but *what* we should do and not do.

These are legitimate subjects for democratic debates, which should be based as much as possible on rational, unbiased arguments. Unfortunately, the issues are so scientifically complex, politically loaded, financially weighty, and emotionally charged that experts with equally impeccable credentials can readily be marshalled on either side of the discussions. Settlement by reasonable consensus is not yet within easy reach. More likely, biotechnologies will, like other innovations of the past, move empirically to progressively wider adoption by the self-improving and self-correcting process that goes ahead wherever attitudes are more permissive or adventurous and thus can gain experience from whatever mistakes or accidents befall these explorations. This will probably happen also in the human sphere. But, here, the ethical concerns are much more serious.

Is the Human Genome Inviolable?

For understandable reasons, applications of the new technologies to human beings have so far been limited largely to nontransforming procedures. Genetic engineering of human subjects is obviously not something to be lightly attempted.

GENETIC TESTING

The simplest human application of biotechnologies has already been mentioned. It is the identification method by way of *genetic imprints*, which does not even require knowledge of sequences and is expected, as we have seen, to replace the old fingerprint method introduced more than a century ago by the Frenchman Alphonse Bertillon.

With the spectacular advances carried out in the last few years in the sequencing of the human genome, which is now close to totally elucidated, much progress has been made in understanding human genetic defects. Thanks to collaboration between clinical geneticists and molecular biologists, mutations responsible for many *hereditary deficiencies* have been pinpointed, leading in turn to the possibility of identifying the bearers of defective genes. Such knowledge can be useful in a number of cases, but may also create problems.

Take, as example, a congenital disease known by the names of two physicians who described it about a hundred years ago, the British ophthalmologist Warren Tay and the American neurologist Bernard Sachs. It is a rare condition, with an abnormally high frequency in certain segments of the Jewish population (who trace back their ancestry to a small central-European village, where the responsible mutation presumably occurred some time in the Middle Ages). Babies born with this defect rapidly develop grave motor and mental deterioration and usually die before they reach the age of three years. Characteristic of Tay-Sachs disease are abnormal lipid deposits in brain and other cells. It is one among many hereditary conditions, known as storage diseases, whose pathogeny has been elucidated thanks to discoveries made by my former Belgian coworker Henri-Géry Hers on another storage disease, in which the abnormal deposits consist of glycogen, a starch-like substance. In these storage diseases, of which more than 50 kinds are known, the deficiency affects one of the digestive enzymes acting in lysosomes (the cell's stomachs).[7] Because of the genetic abnormality, the substances requiring the defective enzyme for their digestion fail to be broken down and thus pile up in the lysosomes, which swell and literally choke their host cells to death, especially in organs such as the brain, where physical constraints oppose cellular expansion.

In genetic storage diseases, as in many other hereditary conditions, the responsible mutations are recessive; they are silent and harmless as long as a normal copy of the gene is present, but they become deadly when each of the parents happens to be a bearer of the same mutated gene and each contributes it to a fertilized egg—the probability is 25 percent—so that the resulting child has two altered copies of the gene. Thanks to the advances of molecular biology, such abnormal genes can now be detected in symptom-free bearers and in fetuses, thus allowing genetic counseling of prospective parents and, if legally permitted and morally acceptable to the parents, abortion of fetuses with two copies of the defective gene.

Diseases of this sort illustrate the advantages—and disadvantages—of diploidy, that is, the fact that genes are generally present in two copies in the genome.[8] If a mutation is recessive, it spares its victim but may thus spread freely, with its only check being the rare cases—more frequent in inbred populations—where two abnormal copies of the same gene join in the genome of a conceived child. If, on the other hand, a deleterious mutation is dominant, that is, is expressed in the presence of a normal copy of the gene, it tends to be self-eliminating unless its lethal consequences become manifest late enough to allow the carrier to produce progeny (of which half, on average, will then bear the abnormal gene).

A dramatic instance of such a case is Huntington's chorea, a fatal genetic disease first described in 1872 by an obscure American physician, George Huntington, who published only that one paper.[9] Huntington's disease is characterized by increasingly disordered muscular seizures, accompanied by progressive psychic degradation and ending in complete dementia. Contrary to most genetic diseases, this particularly nasty condition shows a late onset, at some 40 to 50 years of age, leaving plenty of time for prior hereditary transmission, and it is due to a dominant gene, active even in the presence of a normal copy of the gene. The abnormal gene is known and can be detected, which means that a bearer of the defect can be informed many years in advance of the horrible end that awaits him or her. A well-known American medical personality, Nancy Wexler, whose maternal grandfather, mother, and three uncles died of the disease, has described the drama this situation has created for her and her sister.[10]

In the examples so far mentioned, drastic consequences result from genetic modifications. Such is not always the case. Among the different varieties (alleles) of the same gene present in the human gene pool, a gradation in harmfulness exists. Not all are so maimed that they are unable to perform their function. Some are seemingly normal, except that they augment the risk an individual has of falling victim to a given multifactorial disease, such as atherosclerosis, diabetes, arthritis, or cancer. This knowledge is of great biological and medical interest. But it also generates unforeseen complications.

Who, besides the patients (assuming they have the right to be told), should be allowed access to the information? Should relatives and, if so, to which level of relationship? Should hospitals, insurance companies, prospective employers, justice enforcement bodies, society in general? Should any person or agency—for example, an airline before hiring a pilot—be given the right to order a genetic test on an applicant seeking employment? Furthermore, with the use of computers and their increasing centralization, how is whatever confidentiality one wishes to preserve to be protected? All these problems arose as soon as genetic testing for certain severe diseases became possible. The case of Huntington's disease has been documented in detail by Nancy Wexler. It now appears that the problem concerns everyone, as all of us are carriers of *propensity genes* that increase our risk of getting certain diseases. It will soon be possible to grade every individual in terms of *genetic risk*, creating a nightmarish situation for ethicists in the future.

GENETIC INTERVENTIONS

Genetic discrimination is only one of the many societal problems created by the advances of modern biology. The most far-reaching ones come from genetic engineering. We have seen how this technology has revolutionized biology and provided the means to create new plant and animal species more or less at will. There is no reason, at least technically, why such procedures could not be applied to human subjects. This has become a major issue for the new discipline of *bioethics*.

An important distinction must be made, depending on whether one proposes to act on cells of the *germ line*, thus with possible hereditary transmission of the introduced modification, or on other (*somatic*) cells of the body. For the latter, there is fairly general agreement on allowing interventions of proven usefulness and innocuity. Among these are attempts to correct congenital defects by *gene replacement therapy*. Several children genetically unable to make an enzyme necessary for normal functioning of the immune system, and condemned, for this reason, to life imprisonment inside a sterile bubble, now live an almost normal life thanks to this therapy. Cells taken from their bone marrow or blood were programmed by gene transfer to synthesize the missing enzyme and were subsequently reinjected to produce a population of cells capable of doing the job in the body. So far, not many therapies of this type have been applied successfully because the cells one wishes to correct (for example, brain cells) are not accessible for gene replacement, or because the corrected cells do not survive long enough, or for some other reason. But progress continues to be made in this field, which holds great promise for the future.

Germ-line genetic interventions raise more controversial issues, concerned not only with what is contemplated but also with how it is to be done. In a social climate where even contraception, let alone abortion, still gives rise to heated debates, the prospect of interfering with the human reproductive process could not but provoke strong opposition. In fact, were it not for a ten-year-long effort by two British investigators, the biologist Robert Edwards and the gynecologist Patrick Steptoe, to satisfy the respectable and legitimate desire for a child of couples unable to conceive the "normal" way, the specter of transgenic humans would hardly be on the horizon today. On 25 July 1978, Louise Joy Brown was born in Oldham, England, the first *test-tube baby*, conceived without sexual intercourse. The new era of what the American reproductive biologist Lee Silver has called "reprogenetics"[11] was launched. To many, this means the lid of Pandora's box was lifted.

As soon as relatively safe and effective *in vitro* fertilization (IVF, see above) procedures became available for humans, the new technologies spread rapidly. They are now practiced in a large number of clinical centers all over the world, including some, like my own university hospital in Belgium, of typical Catholic inspiration. It is estimated that more than 150,000 test-tube babies had already been produced by the end of 1994. Their number probably exceeds half a million at the time this book is being written. This success has not gone without problems, such as multiple births, the status and disposal of supernumerary embryos, their use for research, and, in a more bizarre vein, the desire of a widow to conceive with the help of her late husband's frozen sperm or the right of frozen embryos to develop so that they can inherit the fortune of their dead parents. On the whole, however, the benefits of IVF, as a means of helping sterile couples to produce the children they want, are generally viewed as outweighing its drawbacks.

The true ethical minefield created by this technology lies in the "dish phase," during which egg cells and early embryos are accessible to manipulations. Simplest among these is genetic testing for certain traits, which may lead to rejection if the trait is undesirable or, alternatively, to selection of the egg for implantation if the trait is desirable. At least one case is known of parents of a child afflicted with a rare and deadly inborn disease who used this method to produce a new child guaranteed to be unaffected by the disease and whose cells could be used to correct the genetic defect in the older sibling.

Particularly disquieting is the possibility, opened by the new technology, of subjecting human embryos to genetic modifications of a kind now widely carried out on animals, destined to affect all the cells in the body, including germ cells, and therefore to be passed on from generation to generation. So far, the dominant tendency is to proscribe such interventions under any circumstance, on the grounds that they infringe on the *integrity of the human genome*. But there are dissenting opinions. After long deliberations, the International Bioethics Committee created by UNESCO did not, contrary to the desire of many of its members, explicitly proclaim the inviolability of the human genome nor condemn germ-line interventions, in the "Universal Declaration on the Human Genome and Human Rights" that was proposed to the member states, and approved by them, in November 1997. In fact, this document does not address the problem at all, merely demanding "rigorous and prior assessment of the potential risks and benefits" for all "research, treatment or diagnosis affecting an individual's genome." No statement could be more noncommittal.

As of now, possible derogations to the inviolability principle are already envisaged. Should it prove feasible, for example, to extirpate from a transmissible genome the mutant gene responsible for Huntington's disease, there are not many who would oppose the eradication of such an odious gene, one that is surely without any redeeming value. The possibility of introducing desirable genes into germ cells raises much more complex problems. These will be considered in the next chapter.

HUMAN CLONING

With the birth of Dolly, a new, burning problem was added to the concerns of bioethicists. Should human cloning be banned outright? Or should it be allowed under certain conditions? In general, the response—

including that of the UNESCO committee—has been in favor of unconditional prohibition. But this solid wall of opposition is already showing signs of breaches. Projects for commercializing human cloning are even being contemplated.[12]

The desire of egocentric individuals to perpetuate themselves by cloning, the subject of several works of fiction, will probably not be the main motivation behind such attempts. It would, in any case, be based on a misconception. Such a cloning product would be much less than an identical twin of his or her "father" (which could also be a second "mother"). It would be a child of its own times and environment, an authentic individual in its own right, having shared none of the experiences of its so-called twin. But other possibilities have been envisaged. For example, spermless men could father a child by cloning. It would even be possible for lesbian couples to produce offspring by this technique. Another possible application would be "replacement" of a dead person, a procedure sometimes called "missiplicity," from the name of a dog named Missy, whose wealthy owner, in 1998, offered a Texan scientist 2.3 million dollars to clone his beloved pet by nuclear transfer. There have been reports of parents wishing to do the same for their dead child, apparently not realizing the psychological burden such a motive would impose upon the "surrogate" child.

There is, however, another possible application of cloning technology of more promising value. In this form—called *therapeutic cloning*, as opposed to *reproductive cloning*—the embryo is not allowed to develop, but is used as a source of pluripotent cells (*stem cells*) that, thanks to techniques that are being developed, could be cultured to differentiate *in vitro* into, for example, skin, brain, or liver cells. Such cells would be entirely compatible immunologically with those of the donor of the nucleus. One can readily see the enormous prospects for such procedures. The search for compatible donors that so often acts as a bottleneck in successful organ transplantation would be avoided. Instead, each individual could be supplied with a store of compatible cells usable for transplants in the future. Even this procedure, however, remains abhorrent to those ethicists who view any human embryo as a person and consider its downgrading to the status of a "spare parts" store unacceptable.

For this reason—and also because human cloning may not yet be in the offing and is likely to be very expensive if and when it becomes available—scientists are now envisaging the creation of stem cell "banks" from aborted fetuses or from discarded IVF products. Objections have, however, been raised, in the United States and elsewhere, against carrying out the necessary research on human embryos. To obviate these objections, the isolation of stem cells from normal tissues is now actively being investigated. Such cells have long been known to exist and have, in some cases, been used in replacement therapies. Bone marrow, for example, which contains stem cells serving for the replacement of all blood cells, is used on a large scale for therapeutic uses. Present research aims at isolating for the same purpose stem cells that could be used for the repair of other organs. A particularly intriguing possibility, already being considered, is to isolate stem cells from cord blood at birth and to keep them frozen as a personal reserve for the use of the individual later in life.

CHANGING LIFE ITSELF

Just as this book was getting ready for production, news came from Japan that investigators had succeeded in changing the basic language of life.[13] Highlighted by a leading article in the March 2, 2002, issue of the French daily *Le Monde*, this work has drawn the attention of the public to what has already become an important enterprise, involving a number of laboratories in the United States, Japan, France, and elsewhere. The aim of the enterprise is to expand the genetic code and thereby create new proteins and, even, new organisms.

Explained in simple terms, what the Japanese investigators did, building on this international effort, was to introduce artificially into DNA two synthetic, mutually complementary bases, designated S and Y. DNA molecules modified in this manner formed normal double helices, with S pairing with Y, as expected. Through this operation, two new letters (S and Y) were thus added to the four-letter nucleic-acid alphabet and the number of possible codons in the genetic code was expanded from 64 (4^3) to 216 (6^3). Let some of the extra codons specify new amino acids, and an almost unlimited number of new proteins with a vastly expanded range of properties becomes possible. The workers have taken a first step in that direction by engineering *in vitro* transcription and translation systems to incorporate an unnatural amino acid, chlorotyrosine, into a protein.[14]

The result is still a modest one. But the prospects uncovered by this experiment, which is likely to be remembered as historic, are tremendous and almost beyond imagination. What it means is that the way has been opened toward the creation of organisms unlike any organism that has ever existed or could possibly exist within the framework of life as we know it.

The new technologies have already spawned commercial ventures. Their proponents see them essentially as new tools for the chemical industry and insist on their great potential practical benefits. New "bioplastics" with all kinds of desirable physical, chemical, or biological properties will be manufactured. Enzymes will be made that catalyze all sorts of chemical reactions that do not take place naturally. And so on. How society will react to this new development remains to be seen. Judging from the heat generated by the simple transfer of a bacterial gene into a plant, the present possibility of creating new forms of life with new building blocks is bound to be greeted by fierce opposition and accusations of "playing God."

16. After Us, What?

FROM TIME IMMEMORIAL, PROPHETS ANNOUNCE THE IMMINENT end of the world, most often to strengthen their hold on their followers, on the pretext of preparing them for the ultimate ordeal. The prediction is not totally false. According to cosmologists, a time will come when the Sun, its hydrogen stores exhausted, will blow up into a red giant, engulfing Earth in a fiery cloud. But this final blaze is not for tomorrow. Earth has some five billion years left before it goes up in flames. Even though it may not be able to harbor life during all that time, it should be able to do so for at least 1.5 billion years, according to present estimates.[1] This is an enormous amount of time, more than twice the known history of animal life, some 250 times the duration it took a chimp-like primate to become a human being. What may happen during such a long time totally exceeds our imagination. But we can, at least, define certain possibilities. We may even, on the strength of what we know of the past, hazard some predictions.

EARTH LIFE WILL NOT DIE BEFORE EARTH

Life scoffs at the cataclysms that strike the planet that houses it. It even thrives on them. The history of life is landmarked by cosmic, geological, and climatic upheavals that exterminated up to 90 percent of existing species. Each mass extinction was followed, not only by a flourishing of the species that were spared, but even by a creative upsurge of new living forms. Contrary to the "nuclear winter" and other catastrophe scenarios with which the new Nostradamuses threaten us, few circumstances, barring the terminal flareup of the Sun, can be imagined that could erase all life from the planet. Only a total or near-total depletion of liquid water, as may have happened on Mars (see next chapter), could leave Earth lifeless.

Whereas it is likely that life will subsist, the same is not necessarily true of humankind. The normal fate of species is to disappear and give place to others. There are exceptions. Horseshoe crabs have been around, almost unchanged, for more than 200 million years. But such instances are rare. The majority of fossils are of extinct species, to which must be added all those of which vestiges have not yet been found, together with the many "missing links" that have disappeared without leaving any traces. If the human species conforms to the norm, its days are numbered. According to an estimate by the American astrophysicist Richard Gott, if the human species is like other species, it should have no more than a five percent chance of surviving longer than eight million years, not much of a spell in comparison with the span of at least 1.5 billion years left to life. However, there are many who consider calculations of this sort little more than numerology. In any case, the human species is not like others.

The Human Species Holds Its Future in Its Hands

Ever since our distant ancestors began chipping stones and mastering fire, and, especially, ceased their nomadic wanderings in search of food and started to raise their sustenance locally, the face of Earth and of the thin living pellicle that surrounds it has been changing at what has become a truly vertiginous pace. Measured first in millennia, then in centuries, and now in decades, this change has steadily accelerated thanks to the spectacular advances of the sciences and technologies. The world is entirely different from what it was only one hundred years ago, let alone in the times of the Greeks or Romans or, even more so, at the time of Cro-Magnon man. And yet, we are not very different, biologically, from the humans who lived in those days. *Cultural evolution*, not biological evolution, has transformed the world and upset its natural balances.

Many anthropologists have drawn a parallel between the two phenomena (see Chapter 13). Like biological evolution, cultural evolution has been driven by an essentially blind process of natural selection, but acting on cultural rather than on genetic innovations. Qualitatively, the outcome has been the same as in biological evolution: the reproductive success of groups that derived an advantage from their innovations. But quantitatively, this outcome has been dramatically faster and more important than all the preceding steps in evolution. The human species has, in the span of a few centuries, attained an inordinate, almost obscene degree of development. Veritable cancer in the flesh of the biosphere, it covers the planet with ever denser throngs that are exerting an increasing pressure, already close to unbearable, on other living species, natural resources, the environment, and their own internal dynamics.

Until recently, humans did not in the least worry about the consequences of their success. They saw it as the natural and legitimate fruit of their "superiority." They had no other concern than to increase their mastery over nature and spent a good part of their efforts fighting for their "piece of the pie." Savage conquest, colonization, and expansion were the norm. Progress was a goal in itself. Knowledge had no other value than as a source of power. Religion, the only force that could have been capable of tempering this unleashing, served as a pretext for it more often than restraining it.

Only in the last decades has humankind become conscious of the harm its uncontrolled development has inflicted upon nature and of the responsibilities that parallel the exorbitant powers it has gained. It is worth pointing out that it is largely to science, so often maligned by environmentalists, that we owe this new awareness. Thanks to biology, we understand better the functioning of life and the mechanisms of evolution, and we measure more justly the consequences of actions that were not questioned in the past. Moreover, the progress of computer science has given the planet a sort of superbrain that allows us to collect and process a huge amount of information and to diffuse it almost instantaneously to the four corners of the globe. Various world organizations have been created for the purpose of preventing conflicts and coordinating initiatives. This is a radically new situation in the history of humanity. Our future and that of the world of which we are a part are ours to shape, knowingly and collectively, with increasingly powerful means. Such being the case, we have the choice among three possibilities.

FIRST OPTION: HUMANKIND DISAPPEARS

There is nothing impossible about this. When such an eventuality is raised, we think most often of a brutal extinction, such as could result from an encounter with a large comet or asteroid or from a nuclear conflict "to the finish." But such threats are probably not those to be most dreaded. The extraterrestrial threat will be "seen coming" years, if not centuries, in advance, and it will be possible to take measures to destroy or deflect it with the help of nuclear projectiles or other means.[2] If that proves impracticable, it should at least be possible to protect a sufficient part of our biological and cultural heritage to enable the survivors to build a new world on the ruins. As for a nuclear holocaust, we seem to be moving away from it politically,[3] and, in any case, there is little risk that it could be so murderous as to cause the extermination of humankind. But there are slower and more insidious modes of extinction against which we are far from being guarded.

Countless civilizations have disappeared, to the point that many historians, such as the German Oswald Spengler or the Englishman Arnold Toynbee, have seen in the fall of civilizations an almost inevitable consequence of their rise. In the past, this cyclic process has been mostly creative rather than destructive, as there was always, somewhere in the world, an ascending civilization ready to seize the torch, often by benefitting from the acquisitions of its predecessor. But, with the present trend toward globalization, such a process will soon become impossible, as no other civilization will be left to start a new rise.

As of now, the dominant civilization is that known as Western. Born in the Middle East, it has developed by successive cycles, each time widening its territory, which now includes the whole of Europe and the Americas and increasingly extends to the other continents. Heir to a glorious past, Western civilization has produced an extraordinary crop of cultural achievements, including, as its most recent fruit, a remarkable enrichment in our understanding of nature and in our means of acting on it. In spite of its triumphs, one may wonder whether our civilization is still in the ascending curve or has already started on its decline.

There are many, especially among the older generations, who would be inclined to opt for the second possibility, pointing to what they see as signs of decadence in our civilization. Comparisons are often made with the last days of the Roman Empire, increasingly dominated by futility. It is, however, difficult to judge one's own epoch, especially when the pace of change is as fast as it has become today. What is viewed as alarming in the light of the past could be a ferment of novelty and creativity that is about to vivify the future. Or it could just be a transient accident, a tiny fluctuation in the ascending curve of humanity.

Even if our civilization should be truly on the decline, there still remains, in some parts of the world, a reservoir of wisdom that could take over its positive aspects without necessarily adopting its excesses. The day is near, however, when only a single world civilization will exist. What will happen then if it, too, suffers the decadence that historians tell us inevitably follows upon grandeur? Could civilization not just slowly die out, so insidiously as to completely escape the notice of the participants, from a progressive atrophy of its faculties, a sort of societal Alzheimer's disease that would drag humankind down with it to extinction? An imperceptible deterioration of the ability to read and use rational thought— could it have started already?—might well suffice to bring about a slow demise of this sort.

Another possibility that cannot be ruled out is that human folly prevails over all the warnings that caution humankind against the nefarious consequences of its excesses. In such a scenario, the planet's resources slowly fall to exhaustion; deserts continue their inexorable expansion; atmospheric balances are increasingly perturbed; the ozone layer no longer protects life against ultraviolet radiation; the climate warms to the point of causing polar ice sheets to melt, drowning wide coastal fringes; violent storms disrupt the fragile technological network on which the survival of large cities depends; polluted lands and oceans become depleted of plant and animal species. All of a sudden, almost unnoticed and too late for any intervention, the process is irreversible. The planet, victim of the uncontrolled success of the human species, is no longer able to support this species, itself the victim of an evolutionary course that has favored intelligence and inventiveness but failed to provide the wisdom indispensable for the constructive exercise of those faculties.

There is also the possibility that catastrophic epidemics may precipitate the extinction of our species. Unthinkable not so long ago, so great was our faith in the all-powerful means of modern medical science, this eventuality has now become real. AIDS, the Ebola virus, "superbugs" resistant to all antibiotics are here to remind us that medicine has limits and may even, by its advances, paradoxically create diseases against which it is powerless. Other advances, of transportation, for example, render illusory any hope of containing the spread of an epidemic by the traditional means of isolation.

All this goes to show that the human species could very well disappear some day, perhaps distant at our human scale but near in the context of the immense time left for life on Earth. What could happen in such an event? The answer to this question depends on the state in which in our species leaves Earth and the living world.

In the most favorable case, evolution will pursue, with only a brief pause, its dual course toward complexity and diversity. One day, perhaps, a new thinking species will emerge, from a bonobo, a sea otter, a dolphin, or some other species of which we can have no idea today. These successors to *Homo sapiens* will have to repeat our whole journey toward knowledge, as they will find no fossil trace of our discoveries, our theories, or our equations, long dissolved into oblivion. But, differently equipped, they will perhaps arrive at a better view of the Universe, drive their analysis further, and show more wisdom in the application of their findings. In the history of the world they will write, the human species will be no more than an incident among many others, one of those innumerable dead ends of evolution of which some vestiges spared by time will perhaps allow them to recognize one or the other attribute. They will speak of the Science Age as we speak of the Stone Age, and of us as we speak of Neanderthal man.

At the other extreme of possibilities, Earth and the living world will be handicapped to such an extent at the time of our disappearance that only bacteria, which we have seen are almost indestructible, will be capable of long-term survival. Will they be able to reinitiate a new evolutionary process toward increasing complexity? Or will they do no more than cling to some rare, favorable ecological niches, perhaps finally to disappear completely if conditions continue to deteriorate? This is obviously impossible to predict.

SECOND OPTION: HUMANKIND PERSISTS UNCHANGED

For most of us, things could not be otherwise. Even though we may occasionally evoke, as I have just done, the possibility of extinction of the human species, that of its further evolution is rarely envisaged, even in the framework of the evolutionary process from which we are issued.

We reason and behave as though human evolution has stopped with us, as though we have reached a plateau our descendants are destined to occupy indefinitely. To be sure, humans have always lived in this way, within the context of a static and anthropocentric vision of the Universe. But what are we to think now we know that we are the product of an evolutionary process?

Here, a distinction must be made between evolution and speciation. It is generally agreed that the creation of a new bifurcation in the human genealogical tree that would lead to the branching of a new species besides the existing one is not likely. This could happen only if an inbreeding group should be geographically isolated from the rest of humanity for a sufficient span of time—probably many millennia—to reach a stage where crossbreeding with the others no longer gives rise to fertile offspring.[4] The modern world no longer provides conditions where such isolation is naturally possible. As to its enforcement by "ethnic purification," attempts in this direction have been made in the past—Hitler's is the most infamous—and are still being made in some parts of the world, but are not likely ever to prevail long enough against the powerful moral and, even, physical forces that favor mingling. The story of *Homo sapiens sapiens* leaving *Homo sapiens neandertalensis* in the lurch will not be repeated.

This, at least, is what I thought until I came upon *Remaking Eden*, a book by the American Lee Silver, already mentioned in the preceding chapter.[5] In this work, the possibility of speciation by *economic isolation* is evoked, the rich being the only ones able to afford the expensive genetic engineering interventions that give rise to more advanced human beings. I shall come back to this intriguing and disquieting suggestion.

Whatever the future, one thing is certain: *Homo sapiens sapiens* will not remain indefinitely unchanged. There is every reason to believe that this cannot be. All kinds of influences are at work within the human species to modify its gene pool and to provide natural selection with new variants for screening. The question, therefore, is not whether we will change—there can be no doubt about that—but how fast we will change and, especially, in what direction.

THIRD OPTION: HUMANKIND EVOLVES

An important point to be considered here is that the rules of the game have changed with the advent of humankind. Humans can, more than any other beings, affect biological evolution, including their own, by their behavior. They are the only beings able to free this process from its subservience to blind chance and natural selection.

For a number of millennia, humans have affected their own evolution unconsciously, through a sort of interplay between cultural and biological evolution. Take, as a simple example, tool making. Presumably, some individuals were genetically more skillful than others in chipping and shaping stones for specific purposes. To the extent that this ability was selectively useful, the genes responsible for it would tend to spread within the population. Thus a particular biological trait was favored by a cultural innovation.

The recent advances of the sciences, especially of medicine, have considerably increased the influence of the cultural on the biological. In the past, inborn resistance to plague, smallpox, cholera, tuberculosis, or other deadly infections presumably played an important role in natural selection. It even happened that the saving genetic trait was harmful in itself, but less so than the danger against which it protected. A typical example is sickle cell anemia, which was allowed to spread in large parts of Africa because the red blood cells of carriers of the abnormal gene are resistant to the malarial parasite. Today, thanks to sanitation, vaccines, antibiotics, and other means, genetic resistance to infectious diseases has become much less of a selective advantage. For descendants of African bearers of the abnormal sickle cell gene living in malaria-free regions, this trait no longer is of benefit; it has become a severe liability.

Medical progress also tends to modify the composition of the human gene pool in more specific ways, mostly in a direction that counters the normal effect of natural selection. This happens whenever successful treatment allows bearers of harmful genes to reach reproductive age and to produce offspring possessing the same abnormal gene. The discovery of insulin, for example, acclaimed as the miracle cure against diabetes when it was first isolated in 1922, has had as a negative consequence that genes that increase the risk of suffering from diabetes are allowed to spread in the human gene pool. Many other propensity genes (see preceding chapter) are similarly sustained by medical progress.

All these changes in the human gene pool are to some extent incidental; they occur as unwanted side effects of well-intentioned efforts to help individuals. Much more far-reaching consequences arise from our ability to modify genes knowingly and deliberately. We have seen in the preceding chapter how the new technologies have revolutionized biology and given us the means to create new plant and animal species more or less at will. Nothing prevents us, at least technically, from using these means to take our own genetic future into our hands. This has become a major issue.

DESIGNING THE HUMANS OF THE FUTURE?

Our newly awakened awareness of a planetary responsibility has spawned many concerns, debates, and initiatives all over the world, dealing with urgent environmental and societal problems calling for immediate political intervention. Leaving those problems to the experts in charge, I shall, in the following pages, restrict myself to the more delicate—and controversial—question of what interventions, if any, humankind should be allowed, or encouraged, to perform in order to shape itself as a species.

THE HUMAN FAMILY

We are all part of one big family, which can be traced back through the female line to a single "mother of humankind" who lived somewhere in Africa roughly 200,000 years ago. This, with a change in time scale, sounds much like the biblical account, except that our common ancestor was not, like Eve, the sole female in the Garden of Eden. She was one among some 5,000 female congeners living at the time, but she is the only one to have produced an uninterrupted female lineage extending from daughter to daughter unto this day. All the other lineages have died out one after the other for lack of female progeny. How can we know this?

The answer: through the mitochondrial genome. We have seen (Chapter 10) that mitochondria have retained a few genes from their endosymbiotic bacterial ancestors. We also know (Chapter 11) that the maternal egg cell contributes the bulk of the cytoplasm to the fertilized egg. The sperm cell contributes little more than its nuclear chromosomes. Thus, your mitochondria are descendants of your mother's mitochondria, which come from her mother's, and so on. In research originally performed in the 1980s at the University of California at Berkeley by a group led by the late Alan Wilson, samples of human mitochondrial DNA collected in various parts of the world were analyzed and the results used to construct a phylogenetic tree according to the principles outlined in Chapter 7. The tree was traced to a *single root*, often referred to as "mitochondrial

Eve," dated about 200,000 years ago and located in Africa. First disputed on technical grounds, these findings have been confirmed by subsequent work and reinforced by similar male-line studies involving the Y chromosome, which is specific for males (see Chapter 11). The timing is still somewhat uncertain, but the single ancestry and its African origin are generally accepted.

Out of Africa, some descendants of mitochondrial Eve migrated into large parts of Europe and Asia, and, later, Oceania and the Americas, by pathways anthropologists are still trying to trace. In so doing, the early humans fragmented into distinct groups, which evolved separately for a sufficient length of time to develop a number of genetic characteristics adapted to their local environments, giving rise to what used to be called the human races, better named *ethnic groups*. As an example, skin pigmentation, which was most likely dark in the original ancestors and remained so in Africa, where it provided good protection against strong sunlight, may have been largely lost in colder, northern climates, where a pale sun, rather than being hazardous, played an essential role in the ultraviolet-dependent formation of vitamin D in the skin. Other physical and, perhaps, psychological traits developed in the same way as distinguishing marks. In no case, however, was the isolation sufficient for cross-breeding to result in infertile progeny. Incidentally, how the Neanderthals, who probably became extinct only about 35,000 years ago, figure in this story raises an intriguing—and unsolved—problem.

The "races" began to discover each other with the progress of land travel and navigation. Certain common traits stood out, besides the many obvious differences. All groups, whatever their degree of development, or lack of it, had a language, sometimes highly elaborate in spite of a stone age technology. All practiced various forms of art. All made tools and took care of their sick according to methods that were transmitted from generation to generation. All had a lore, a cultural tradition, an organized social structure, a set of myths and religious rites. All buried or otherwise attended to their dead, often in a manner destined to ease passage to a hypothetical other world. In other words, all, beyond their differences, were fundamentally *human*, as confirmed by their ability to interbreed and, now, by their mitochondrial DNA.

Understandably, this equality was not readily recognized. The "white race," in particular, arrogantly influenced by its technical and cultural "superiority," did not hesitate to affirm its dominance over the others in ways that, less than 150 years ago, still included slavery. Such attitudes are now condemned by most societies, at least legally. But they remain deeply imbedded in the mentality of many, as evidenced by the continuing success of racism in many parts of the world. Few would still support active discrimination and exploitation of "lower races," but protection of a "superior race" by various means is still on the agenda of many extremist groups. This kind of selection has long been imposed more or less as a matter of course by the many geographical, social, religious, and other factors that have favored—and still do—unions between persons belonging to the same ethnic group. The idea of using science to deliberately promote this goal is recent.

This effort was launched in Victorian England, under the name "eugenics," by a cousin of Darwin, the physiologist and anthropologist Francis Galton. The eugenic agenda has been promoted by many famous scientists, including several Nobel laureates, among them the Frenchman Alexis Carrel, the Americans Hermann Müller and William Shockley, and the Austrian Konrad Lorenz, who went so far as to approve Hitler's racism. Indeed, in addition to the innumerable crimes they perpetrated in the name of ethnic purification, the Nazis recommended the positive selection of a "*Herrenvolk*" through unions between "purebred Aryans."

Because of these and other horrors, eugenics has become a dirty word in our societies, not just politically incorrect but truly banished from any decent person's vocabulary. Yet, we must reconsider it in a new version, because modern science has now given us the means to actively and deliberately manipulate the human genome. Should we—or can we—ban such interventions? If not, how should we go about it?

THE EUGENICS OF THE FUTURE

Let us be clear about it. What science can do, it will do, some time, somewhere, whatever obstacles may be put in its way. Research itself can be hemmed in by regulations. But, even so, advances deemed undesirable will not be easy to avoid, because of relentless human curiosity and the unpredictability of basic research. Who, in the early 1960s, could have found a valid reason for preventing Arber from trying to understand restriction-modification and thus from carrying out, as seen in the preceding chapter, investigations that turn out, in retrospect, to have made genetic engineering possible? In addition, the very idea of prohibiting research that does no harm in itself, simply because of its possible consequences, is deeply repugnant to scientists, who feel strongly that nothing should stand in the way of the pursuit of truth by legitimate means. Methodologies must be regulated, but objectives should not.

We are not dealing with research, however, in the present case, but with the application to human beings of technologies of proven feasibility. We have seen in the preceding chapter how we are on the verge of being able to genetically modify human germ-line cells. Note that this kind of selection has already long been practiced in a negative way, through abortion of abnormal fetuses and, more recently, through the destruction of abnormal embryos. The question, today, is whether *positive* selection should be allowed and, if so, under what conditions.

There is a strong tendency in support of proscribing all such interventions as infringing on the integrity of the human genome. One may wonder, however, what higher principle justifies such an attitude. As mentioned in the preceding chapter, the International Bioethics Committee of UNESCO refrained, after long deliberations, from proclaiming the inviolability of the human genome. As already pointed out several times in this book, the present composition of the human gene pool is the outcome of several million years of evolution—to consider only the last hominization steps—in the course of which the combined effects of innumerable biological and environmental factors have provided natural selection with the elements from which the present genetic diversity of the human species has issued. The traits thus retained are those that, under certain circumstances of the past, turned out to be useful or at least, not significantly detrimental to the survival and reproduction of the individuals concerned and their progeny. There is no reason to believe that these traits represent an optimal set with respect to present conditions. The case of the sickle-cell gene, mentioned earlier, is a typical example of genetic features, of which there may be many, that were retained thanks to selection factors that do not exist any more today in large parts of the world. According to the sociobiological school, this could be true also of a number of behavioral traits that were useful to our hunter-gatherer ancestors but have ceased to be beneficial in our present societies. The genetic inheritance of humankind includes many museum pieces of no value in the modern world.

There is thus no objective justification for endowing the human genome with a sacred character that would render it inviolable just because it is the product of natural causes. Its protection is in any case an illusory objective, considering the many mutations and other accidental changes beyond our control that continually affect it, as well as the numerous indirect and often unpredictable consequences of human activity. The role of medical treatments has been mentioned earlier. In reality, to protect the human genome does not merely mean trying to preserve what is; it also means abandoning the future to chance. It is difficult to condone such a denial of responsibility, if it is no longer inevitable, unless one adheres to some sort of mystique that deifies nature and puts greater trust in chance than in reason.

There is little doubt that future generations will increasingly interfere with the human genome, not even stopping at the germ line. There will always be exceptions to whatever rule has been edicted, and experience shows that exceptions tend to multiply until they become the norm. The recent history of contraception, abortion, and, lately, euthanasia and assisted suicide is there to prove it. Even human cloning, which met with solemn declarations of disapproval when the birth of Dolly was first announced, is already viewed as a viable option under certain conditions (see preceding chapter).

Genetic engineering of human beings is still dependent on highly sophisticated and expensive technologies that run no risk of mass applications. But, here again, experience shows that what is rare today often becomes commonplace tomorrow. Just think of personal computers, to take a recent example. I shall not see genetic engineering go the same way as personal computers, perhaps not even my children nor my grandchildren will do so. But some future generation will and will be faced with deciding on desirable directions. What its priorities will be cannot be predicted, but one can venture some guesses and formulate some wishes.

First, it is to be hoped that the decision will not be left to some all-powerful bureaucracy reminiscent of the most nefarious dictatorial regimes of the past. To obviate this danger, the British philosopher Jonathan Glover has recommended a "genetic supermarket," where prospective parents could order their children à la carte.[6] This solution has the advantage of avoiding a centralization of powers that might some day put the future of humankind into the hands of a dangerous minority. But it does little more than replace one form of randomness by another. There is no assurance that the individual choices of parents will collectively exert more favorable effects on the human gene pool than the blind play of mutations. What shall be done if there is an excessive demand for baby Mozarts or Michael Jordans?

This question illustrates another difficulty of eugenics, whoever holds the power of decision. Most human abilities one might want to favor, such as intelligence, musical ability, or athletic performance, depend on highly complex genetic and environmental influences that are still far from being understood. At present, one of the most controversial and fiercely debated issues concerns the nature-nurture problem, with, at one end of the spectrum, those who believe that "it is all in our genes," and, at the other, those who claim that the environment does it all. Before humankind reaches a state in which it can direct its own evolution, enormous advances will first have to be made in our understanding of the relationships between genes and hereditary characters, especially at the level of the mind.

Upon reflection, what would seem most desirable as a goal would be to enhance *all our mental faculties* simultaneously and harmoniously. What gives our species its special significance is our intelligence, our sensitivity, our imagination, the whole set of mental abilities that allows us to apprehend the world, to understand our own nature and that of the universe to which we belong, to express what we feel in writings and works of art, to ask questions, to wonder. The royal road of evolution clearly runs by way of those faculties. Enhancing them eugenically is a logical step, which does no more than help evolution along its continuing vertical drive toward complexity. It may even be an obligatory step if this drive is to proceed beyond the human stage, in view of our newly gained mastery over our biological future. What, until now, was driven only by natural selection has to some extent become dissociated from it by our ability to affect the direction of evolution. Furthering mental development thus appears as almost a duty for humankind, as it represents our best hope—if not the only one—to avoid degeneration and, perhaps, extinction.

According to present evidence (Chapter 14), the richness of mental life seems to depend on the surface area of the cerebral cortex; remember Lucy's reverie. If such is the case, then further expansion of the cerebral cortex appears as the most desirable goal of the eugenics of the future. Perhaps one day, enough will be known of the genes that govern the size of this part of the brain so that it will become possible to induce such a process. Even with this kind of knowledge, the objective may still be unattainable without additional, coordinated interventions.

We have seen, in Chapter 12, how the human cerebral cortex has been allowed to expand thanks to a complex process of neoteny, which has caused birth to happen at an increasingly immature stage of brain development. Unless means were found, such as genetically induced widening of the female pelvis or generalized use of cesarean section, to permit greater enlargement of the head *in utero*, further expansion of the human cortex would probably require birth to take place in an even more immature stage than at present, followed by an even longer period of postnatal helplessness and brain maturation. Whatever the problems that may be encountered, it certainly seems plausible that humans may one day, by rationally combining genetic and cultural influences, improve what has so far been achieved under the sole pressure of natural selection.

Granting that such manipulations will one day be feasible and acceptable, who should benefit from them? Obviously, this should be everybody, without discrimination. But there may be a problem, namely that of cost. Today, IVF, as applied to humans, is an expensive procedure, accessible only to those who can afford it. What if this should be the case also of the germ-line interventions of the future? Lee Silver, already cited earlier in this chapter, has raised this possibility to an almost nightmarish situation in which only the wealthy would be able, generation after generation, to improve their genome, up to the point of creating a new species no longer able to crossbreed with the indigent majority left to reproduce in the "normal way." Reproductive isolation, the condition of speciation, would thus be accomplished without geographical isolation.

FIRST THINGS FIRST

The measures just sketched hardly qualify as priorities, since the means to accomplish them are clearly out of our reach and will probably remain so for many years. No doubt, genetic studies and interventions on human beings will continue to proceed slowly and cautiously, mostly for medical purposes and under appropriate safeguards. As progress is made, feasible objectives will gradually clarify and be considered by the bioethical committees of the time. All this is bound to happen in the play-it-by-ear mode. Long-term programming is not possible, especially in a field in which so many unexpected discoveries are likely to be made.

A major problem in planning for the future of humankind is that our ephemeral nature condemns us to live on a time scale that is negligible with respect to that of evolution. Our preoccupations are with our immediate future and that of our children and grandchildren. For political leaders, the span is even shorter, the few years that separate them from the next election. What exceeds half a century is of interest only to a few thinkers almost nobody listens to. We live in the present. How can we face responsibilities that extend over millennia? Fortunately, there are things we can do right away, within the context of our physiological time.

Our first concern should be for our environment, for the simple reason that it raises problems of great urgency directly linked to our immediate future and that of the coming generations. The questions involved are highly complex and are now being debated at many levels of government, both nationally and internationally. I shall not discuss them here, except to express the wish that science and reason be accorded the place they deserve in these deliberations, recommendations, and decisions.

Another pressing problem concerns the size of the human population. There are too many human beings on our planet, and the increase in their number—almost 100 million per year—will soon become intolerable. All our efforts must tend to limit this expansion as rapidly as possible. If we don't do so, natural selection will do it for us, with dramatic consequences whose premonitory signs are already discernible in various parts of the globe. Famines, epidemics, conflicts, mass expulsions, genocides, ethnic purification, terrorist attacks, increasing hostility against immigrants, and dehumanizing urban environments multiply the world over. Although it would be simplistic to impute all those ills just to demographic expansion, this phenomenon certainly plays a dominant role in them. Our privileged ability to replace natural selection commands as our first duty that we stem the growth of the world's human population.

Next to the quantitative problem, there is a qualitative one that requires humankind's urgent attention: the fate of its *children*. At present, not one child in one thousand is allowed to reach the mental development it is genetically entitled to. In Chapter 14, we saw the crucial role of outside influences in the early wiring of the brain. Most of the world's children are deprived of the stimuli necessary to develop the rich neuronal network their genetic potentialities would allow. Many even lack the material requisites of such a process, for want of adequate nourishment.

Education, even at the primary level, is still denied a majority of the young.

These facts should set our immediate priorities. Before even thinking of rendering humankind genetically capable of greater mental performances, let us focus on giving all its members the chance to realize the potential they are born with. Let us first provide suitable economic, social, and family conditions, as well as appropriate educational opportunities, everywhere. Measures of this sort should, by themselves, without calling on genetic interventions, favor hereditary traits that lead humanity to a higher level of mental development and, thereby, render it better able to assume its future responsibilities. Thus, humanism on a human time scale can be inserted into an evolutionism of cosmic dimension. But, for this to happen, humankind will have to unlearn many of the selfish, short-term instincts that permitted its earlier evolution and have continued ruling much of its recent history. Some signs of such a trend are discernible in the present-day world; whether it will prevail in time to avoid major global catastrophes is, however, far from certain.

Up to now, I have considered only ways of improving the capacities of individual brains. But there could be another way of enhancing the mental scope of humankind, namely by putting many brains together. There are indications that this could be happening in front of our eyes, without our being aware of it.

ARE HUMAN SOCIETIES COALESCING INTO A SUPERORGANISM?

Are we going the way of ants and bees, or perhaps even the way the unicellular founders of pluricellular organisms went one billion years ago? There are some who see in the evolution of our societies harbingers of such a transformation. Individual humans have long ceased to be capable of independent existence, relying more and more on mutually complementing skills and competences to achieve, collectively, a level of living no individual could attain alone. Already early communities divided responsibilities among bakers, butchers, carpenters, bricklayers, soldiers, teachers, administrators, and other specialized professionals. This kind of interdependence has done nothing but grow, both in depth and in width, to the point of now forming a highly intricate network that envelops the planet.

There was a time, a few centuries ago, when a single person, like the young Italian Renaissance polymath Giovanni Pico della Mirandola, could boast—in what was obviously a gross overestimate, even at that time—of having mastered the whole of human knowledge. No sensible person, however gifted, could make such a claim today. Available knowledge now fills innumerable written texts, stored in libraries and, increasingly, in electronic memories, to which everyone has access but of which no one can assimilate more than a minute fraction. Experts have become increasingly specialized and their languages increasingly esoteric. Attempting, as I have ventured to do in this book, to draw the "bigger picture" has become "mission impossible."

Human knowledge still exists as a body but is fragmented into a large number of separate parts now integrated only in a computer-supported network of communication. Even the sifting of knowledge is increasingly done by computers. Genetic analysis is a typical example. Gene sequences are stored in huge data banks and handled with the help of sophisticated programs designed, for example, to find "open reading frames," that is, translatable sections framed by an initiation and a stop codon, or to derive the amino acid sequence of a protein from the base sequence of a gene, or, more elaborately, to detect similarities between sequences, assess their degree of kinship, and use the results to construct phylogenetic trees.

Literary and artistic productions are similarly centralized in various kinds of libraries, museums, and collections, for which the Greek word *thêkê* (chest) has provided a universal suffix (especially in French): bibliothèque (*biblion*, book), pinacothèque (*pinax*, painting), glyptothèque (*glyptos*, carved), cinémathèque, discothèque, médiathèque, and so on. Here as well, computers are providing increasing support.

Also in economic, social, and political areas, multinational structures are coordinating and overseeing the operations of governments, legislatures, financial institutions, industries, commercial enterprises, healthcare delivery systems, and other organizations worldwide. Moral responsibilities and ethical concerns likewise have become globalized, in areas such as environmental protection or bioethical safeguards, for example. World organizations and world congresses abound.

So it appears that humankind has become a *superorganism*, composed of multiple organs kept together by a growing network of integrative communications, directed by a veritable superbrain with the assistance of increasingly powerful computers, subject to the commands of a collective conscience that has acquired a planetary dimension. We are approaching the scenario imagined by the American sociologist Gregory Stock in his 1993 opus, *Metaman.*[7]

Where will it end? Is it conceivable that, similarly to the cells of multicellular organisms, human beings will abandon more and more of their autonomy to the centralized dominance of a "plurihuman organism?" Or could it be, as some enthusiasts of "artificial intelligence" would have us believe, that machines will some day not only accomplish performances beyond our abilities, as they obviously already do, but even experience internal, conscious manifestations that apprehend more than we can conceive? We can't know. For my part, I find it difficult to imagine that electronic circuits will ever be able to replace the polyneuronal circuits in their mysterious power to generate conscious thought. In my view, individual conception, imagination, and creativity remain inescapable components of the future and deserve to be fostered in every possible way.

As we ponder the future and look into possible ways of shaping it, one additional element remains for us to consider. What if conscious, intelligent, humanlike beings, perhaps superior to us, exist elsewhere in the universe? What are the chances of our ever finding out about this? And, if ever we can, how would such a discovery affect the fate of humankind? The next chapter is devoted to these questions.

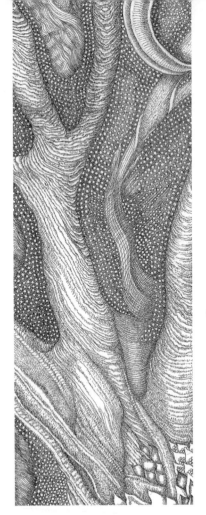

17. Are We Alone?

W HO, GAZING AT THE STAR-
lit sky, has not wondered
whether there is anyone
"out there?" From the philosophers and
theologians of the past to the scientists
and science fiction writers of today, this
question has not ceased firing imagina-
tions, feeding speculations, and fueling
debates; it has now become a legitimate
object for the scientific enterprise.

AN AGE-OLD DREAM

More than 2,000 years ago, the Roman poet and philosopher Lucretius
reasoned that ours cannot be the only inhabited world. "Confess you
must," he wrote in his *De Rerum Natura*, "that other worlds exist in
other regions of the sky, and different tribes of men, kinds of wild beasts."
For agreeing with Lucretius, the Dominican friar Giordano Bruno paid
with his life in 1600, burned at the stake on a Roman piazza, by order
of the Inquisition.

Extraterrestrial civilizations made a dramatic entrance into science in
1877 when the Italian astronomer Giovanni Schiaparelli proposed that
the lines detected on the surface of planet Mars by his colleague Angelo
Secchi were artificial waterways. Intrigued by this notion, a wealthy
American astronomer, Percival Lowell, built a special observatory in
Flagstaff, Arizona, for the sole purpose of studying Mars. He mapped

many of the alleged canals and conjectured how the Martians had dug them to irrigate the surface of the barren planet with water tapped from the poles. This idea, in turn, caught the imagination of the British writer Herbert George Wells, who had the Martians invade Earth in his *War of the Worlds* (1898), which was adapted for radio 40 years later by the American movie director and actor Orson Welles, causing widespread panic when the piece was first broadcast on Halloween eve, 30 October 1938. Mars is still very much in the news today, but no longer as the site of an extraterrestrial civilization, whether friendly or menacing. The Martian waterways indeed exist, but they have been dry for more than three billion years, and they were dug by nature, not by little green men. Whether their banks ever bore life has become a question of burning interest.

Exactly 23 years after the 1938 panic, on Halloween eve 1961, a small group of distinguished scientists, among them the American chemist Melvin Calvin who had just been awarded the Nobel prize in chemistry for his work on photosynthesis, and the charismatic American scientist and television personality Carl Sagan, met at the National Radio Astronomy Observatory, in Green Bank, West Virginia, at the invitation of a young American astronomer, Frank Drake. The purpose of the meeting was to discuss ways of detecting radio signals that might be sent by some extraterrestrial civilization. Remarkably, this outlandish—in both senses of the word—project eventually led to an undertaking of considerable magnitude, supported by the National Aeronautics and Space Administration (NASA) under the acronym SETI (Search for Extra-Terrestrial Intelligence). When a parallel project was started in the Soviet Union by the astrophysicist Iosif Shklovsky, the two superpowers entered into a major international cooperation at a time when *glasnost* was still far off in the future. Today, NASA no longer supports the SETI project, but the search continues with private support, under the direction of astronomer Jill Tarter.

Not all space exploration enthusiasts were ambitious or optimistic enough to look for extraterrestrial intelligence; many would have been quite content with evidence of extraterrestrial life. At present, efforts in this direction are so diverse and important as to have spawned a new discipline variously termed exobiology, bioastronomy, or astrobiology, with its own institutes, meetings, and publications.

In this connection, something of a climax was reached on 7 August

1996, when a special, televised press conference, introduced by President Clinton himself, was convened in Washington by NASA administrator Daniel Goldin, to announce to the world that evidence of past life on Mars had been detected by a team of American investigators led by NASA geologist David McKay. No expensive mission had brought this startling news to Earth, but rather a chance missile, a 1.8-kilogram piece of rock discovered in Antarctica, in December 1984, by Roberta Score and six other American explorers and identified as most likely originating from Mars. Dislodged from the surface of that planet some 16 million years ago by an impacting object, the meteorite, now famous under its classification number ALH 84001, had, after a long sojourn in space, wandered close enough to be caught by Earth's gravitational pull, ending up in Antarctica, where it remained buried in the ice for some 13,000 years until it was dug out, carefully put in a sterile wrap, and transported to the United States for analysis. As I shall mention later, the ALH 84001 evidence has failed to convince a majority of experts. But the enthusiasm for life on Mars and other extraterrestrial sites remains unabated. If we were asked to accept bets as this book is written, what odds would we offer for the existence of life elsewhere in the universe?

THE CASE FOR EXTRATERRESTRIAL LIFE

First, we must decide what we mean by "life." I shall stick here to my definition of Chapter 1, based on the properties of life as we know it, not, of course, necessarily in the form of animals, plants, or even bacteria present on Earth, but built with the same kind of chemistry under the same kind of cellular constraints. One often reads of possible forms of life constructed with molecular components other than proteins, nucleic acids, and other typical biological constituents, or even made of different atoms, for example, with carbon replaced by silicon, its closest kin in the table of elements. There is at present no valid basis for such speculations. In any event, if we are to identify life by chemical or fossil traces, we must perforce draw our criteria from what we know of life as it is.

With this definition in mind, what do we know or suspect of the existence, past or present, of life outside Earth? Let us look at our own solar system first.

LIFE IN THE SOLAR SYSTEM

Our two nearest neighbors, Venus, which is closer to the Sun than Earth, and Mars, which is further away from it, occupy the outer edges of what is sometimes called the *habitable zone*. Of the two, Venus, with a surface temperature close to 500° C, is definitely too hot nowadays to harbor life. It may possibly have been habitable in the early days after its formation. But the question is essentially academic. No doubt, we shall never know.

Mars, on the other hand, with an average surface temperature of −53° C, is too cold to harbor life, at least on its surface. It has a thin atmosphere, made mostly of carbon dioxide, part of which freezes every winter to cover the poles with a white cap, formerly believed to be made of water ice but now identified as what we know as dry ice, the stuff spewed by fire extinguishers. There is abundant water on Mars, however; it exists as permanent ice underneath the North polar cap of dry ice and in the soil in the form of permafrost, such as is found in some parts of Siberia, for example. It is generally agreed that liquid water must be present in some areas below the permafrost blanket. Should molecular hydrogen be available, bacteria similar to some forms present in Earth rocks could possibly exist there. How deep in the Martian crust one would have to dig to find them is, however, unclear.

There is strong evidence that Mars enjoyed a milder climate at a younger age, some four billion years ago. This is indicated by the "canals" mentioned above, which show unmistakable signs of having been carved into the planet's surface by some running liquid, most likely water.[1] This fact suggests that living organisms may have been present on the Martian surface at that time. There is thus an enormous interest in looking for signs of present or past life on the red planet.

The first attempt in this direction was made in the summer of 1976, when two beautifully designed, robotized laboratories landed safely on the surface of Mars, launched from the *Viking* orbiter spaceships. The robots successfully carried out three kinds of chemical tests on samples of Martian soil. At first, the results they sent out to Earth seemed indicative of life. But they proved to be "false positives," due to nonbiological mechanisms, after a fourth test, performed by a highly sensitive technique, revealed no trace of organic carbon.

Enthusiasm for the search for life on Mars was seriously dampened by the negative results of the *Viking* mission, until the ALH 84001 meteorite dramatically reawakened interest in this quest. In this case, all the resources of modern technology could be applied to the rock, allowing highly sophisticated tests. A number of chemical analyses yielded results that, singly, could be considered only suggestive, each being explicable by nonliving mechanisms, but, together, were viewed by the investigators as strongly indicative of past life. What appeared to be the clinching argument was the morphological detection of small grains taken to be the fossilized remains of "nanobacteria" (from the Greek *nanos*, midget, which has become a prefix meaning one-billionth, as in nanometer). Upon critical examination, this evidence turned out to be the weakest of all, as the size of the alleged Martian organisms (some 25 nanometers) is at least one order of magnitude smaller than that of any known bacterium and, in fact, is such that it could not possibly house the strict minimum needed for autonomous Earth life. The matter is still being debated, but most experts are skeptical, agreeing with the American microfossil specialist William Schopf, who, at the 1996 press conference, quoted Carl Sagan as having said: "extraordinary claims require extraordinary evidence," clearly implying that this requirement had not been met.[2]

Whatever the issue, ALH 84001 has had the merit of stimulating new projects. Unfortunately, next to the successful launching of two spacecrafts in 1996, *Global Surveyor* and *Mars Pathfinder*, two catastrophic failures have seriously compromised the program elaborated by NASA. On 23 September 1999, the *Mars Climate Orbiter*, instead of being put in orbit around the planet, was sent crashing to the planet's surface by a command erroneously calculated in American units while the equipment was programmed in metric units. Cost of this mistake no high-school kid would have made: 125 million dollars. Less than three months later, the *Mars Polar Lander*, worth 165 million dollars, lost all contact with Earth after a poorly programmed, overly rough landing. These disasters have slowed down implementation of the program but without causing it to be abandoned. Several missions are planned for the next few years, including at least one in which highly sophisticated, robotized equipment will dig material one meter below the Martian crust and subject it to a number of critical analyses capable of revealing traces of biological substances, such as proteins. On the other hand, the project to send a manned mission has been indefinitely postponed. In this project, a spacecraft had been planned to carry a fully equipped laboratory in which the crew would perform fairly complicated experiments and choose the materials they were to take back to Earth for further analysis.

There is also interest in some celestial objects too distant from the Sun to be sufficently heated by radiation but deriving enough heat from internal sources, such as tidal friction or volcanic activity, to be able to contain liquid water and, therefore, to harbor life. Especially promising are the Jupiter moon Europa, which appears to be covered by ice, most likely surmounting a liquid ocean, and the Saturn moon Titan, which is believed to possess seas of liquid methane and other hydrocarbons, below which there could be water.

The example of Titan illustrates one of the difficulties in ascertaining the existence of extraterrestrial life. The mere presence of organic carbon compounds is not enough proof. We saw in Chapter 3 that organic molecules, including such typical biological constituents as amino acids, are found in many extraterrestrial sites, where their formation is almost certainly due to nonbiological chemical processes. The problem was also encountered with the ALH 84001 meteorite, which was found to contain traces of materials known as PAHs (polycyclic aromatic hydrocarbons). This clue was not considered demonstrative by the critics, because similar substances have been identified in comets and meteorites.

There is another difficulty. Should authentic evidence of life ever be discovered on Mars or on another component of the solar system, the finding would obviously be of tremendous interest, but it would not provide a definitive answer to our main question. The problem would remain whether life actually arose locally or came from Earth, carried by some meteorite. The alternative possibilities that Earth life came from the extraterrestrial site or that both terrestrial and extraterrestrial life came from some third source also must be considered. Given the distances involved, such possibilities cannot be ruled out.

LIFE IN THE GALAXY

Our Sun is but one among some 100 billion stars, arranged in a large, disk-shaped swarm with a diameter of about 100,000 light years, or one billion billion kilometers (light, travelling at 300,000 kilometers per second, would take 100,000 years to cross the disk). When viewed through the plane of the disk, on a clear summer night, this huge cluster of stars appears as a white trail in the sky, the Milky Way, or Galaxy (*galaxias*, from *gala*, milk, is the name the ancient Greeks gave to the Milky Way).

According to modern theories, stars arise through the gravitational collapse of swirling clouds of gas and dust. In this process, the cloud flattens into what is called a *protoplanetary disk*, the heart of which condenses by gravitation into the central star, which heats up tremendously and becomes an active generator of nuclear energy, while the peripheral parts fragment into separately condensing bodies, the planets. Depending on the size of the initial cloud, the star becomes anything between a dwarf and a giant, with significantly different histories. About one-third of the Galaxy's stars have sizes sufficiently comparable to that of the Sun to allow the hypothesis that they have a similar history. If all this is correct, there must be, in our galaxy, billions of solar systems, of which a significant fraction may include a planet with Earthlike properties.

Fortunately, this hypothesis is theoretically accessible to verification, because the Galaxy is in constant evolution, with stars continually being born and dying. Thus solar systems in various stages of their histories should be there to be observed, given sufficiently sensitive techniques. Recent advances are beginning to allow this in practice. Clear evidence of the existence of protoplanetary disks has been found around several nearby forming stars. From present results, it is estimated that between one-quarter and one-half of the young stars in our galaxy have disks around them.

Planets cannot usually be observed directly in the glare of the star, but their presence can be ascertained indirectly by a method that detects the periodic wobbling of the star caused by the gravitational pull of the orbiting companion. There are two problems with this technique. First, it does not readily allow the distinction between a true planet and a "brown dwarf," which is a smaller, companion star that forms by a different mechanism. The second difficulty is that only objects sufficiently large and near their sun to cause a measurable degree of wobbling can be identified in this manner. Our own planet, for example, could not on its own betray its presence in this way; its pull on the Sun is too weak.

The first circumstellar object to be detected by this method, later identified as a brown dwarf, was discovered in 1989 around a star catalogued under the number HD 114762. Since then, more than 50 companions, of which many are considered true planets, have been located around nearby stars. There is also hope, supported by recent observations, that further technical progress will allow direct visualization of the companion.

We are still far from detecting an actual Earthlike planet capable of bearing life. But present results are important in that they indicate that planet formation is a frequent concomitant of star formation. The theoretical surmise that a large number of stars in our galaxy may be surrounded by planets is thereby comforted, making the presence of an Earthlike planet around a significant subset of stars very likely on probabilistic grounds. An earlier estimate that the Galaxy may contain as many as one million planets with a history comparable to that of Earth thus appears plausible. Note, however, that this opinion is far from being unanimously shared.

What proportion of the habitable planets of our galaxy, assuming some exist, actually bear life? It is not likely that this question will ever be answered in concrete terms, as only a minuscule portion of the Galaxy is accessible to direct exploration, even with the most advanced means of space travel conceivable. Only incoming radiation can inform us, and even that information is flawed, since radiation does not travel instantaneously. The news we receive is bound to be stale by up to many tens of millennia, depending on the distance separating us from its source. In addition, for radiation to tell us something about the presence of life, some life-specific signal would be needed. As we have seen, spectral evidence of the presence of organic molecules does not suffice for this purpose. A possibility that has been evoked is to rely on the detection of molecular oxygen (or of oxygen-derived ozone). This would indeed be a strong indication of life, but in an advanced form. Remember that it took Earth life close to two billion years before it started raising the oxygen content of the atmosphere. In any case, whatever the signal adopted, the technical problems of detecting an almost imperceptible emission close to a star's enormously stronger brilliance appear today as totally insurmountable.

This is no reason to be discouraged, however. As pointed out in the early chapters of this book, there are strong reasons to believe that life arose through highly deterministic chemical processes that were bound to take place under the prevailing physical-chemical conditions. We now find that our galaxy probably contains many solar systems, of which a number may include an Earthlike planet on which life-generating conditions could be duplicated. There is thus a significant probability that life may be abundant in our galaxy. If this is true, there could possibly be a life-bearing planet near enough in our neighborhood to allow its detection, perhaps not with present-day technologies, but with those of the future. How many times in history has the impossible of today become the reality of tomorrow! Finding life elsewhere would be such a tremendous discovery that a substantial effort to enable it deserves to be made.

LIFE IN THE UNIVERSE

Until the early part of the twentieth century, we knew of only our galaxy. To be true, some fuzzy objects, called *nebulae* for this reason, had been detected in addition to stars. But only in the 1920s were nebulae clearly identified as galaxies by the American astronomer Edwin Hubble, using a newly built, giant telescope at the Mount Wilson observatory, in California. Hubble made an even more important discovery, known as the "*red shift*." Basically, what he found was that the characteristic wavelength of a given type of radiation received on Earth is shifted to an increasingly higher value the more distant its source. He interpreted this finding in terms of a phenomenon first studied for sound waves by the Austrian physicist Christian Doppler, and later extended to light waves by the French physicist Hippolyte Fizeau.

In the domain of sound, the Doppler effect is familiar to all of us. When a hooting car or train moves toward us, the pitch of the sound goes up progressively (shorter wavelength), to subsequently fall (longer wavelength) once the vehicle has passed us. This is understandable. The waves are compressed as their source approaches us and expanded as the source moves away from us. The Doppler effect thus gives information on the direction (by its sign) and on the speed (by its magnitude) of a moving source of sound. Transposed to the light emitted by a star or galaxy, the Doppler (Fizeau) effect provides the same type of information. It allowed Hubble to conclude that the stars and galaxies move away from us (shift to higher wavelengths), which has become the central piece of evidence supporting the concept of an expanding universe that started from an original Big Bang. In addition, Hubble made the cardinal observation that the velocity with which stars move away from us is directly proportional to their distance from us. On the strength of this relationship, now known as Hubble's law, he was able to infer from the red shift of the nebulae that they were much further away from us than the stars of our galaxy, thus identifying them as distinct galaxies. These monumental achievements have been commemorated in the satellite-borne

Hubble Space Telescope, which is now scanning the skies far above any interference by Earth's atmosphere. It will be remembered that this splendid piece of equipment was almost ruined by a flaw, a distortion of its mirror, which was corrected by one of the most delicate space missions ever undertaken.

Today, more than 100 billion galaxies are known to exist. The nearest ones, the Magellanic Clouds, are between 150,000 aand 200,000 light years away from us, near the edge of our own galaxy. The most distant galaxies are almost 15 billion light years away from us, making their existence known to us by information emitted at the dawn of the universe. Thus from the nearest to the most distant galaxies, we receive a cut through time spanning almost the entire history of the universe. This is a tremendous boon to astronomers and cosmologists. But it is of little help to exobiologists.

If evidence of life is hardly likely to be detectable in our own galaxy, except in our immediate neighborhood, the search for life elsewhere is obviously hopeless. All we can say is that if other galaxies are like our own, they too may be teeming with life. Multiplying the estimate of one million life-bearing planets in our galaxy (see above) by the number of galaxies (on the order of 100 billion), we arrive at the conclusion that there may be as many as 10^{17} (one followed by seventeen zeros) foci of life in the universe. Even allowing a margin of error of many orders of magnitude, we are still left with a respectable number of planets able to give rise to life as we know it. Unless the astronomers' estimate and my own are completely off the mark, *life is widespread* throughout the universe.

The Case for Extraterrestrial Intelligence

Put in simple terms, the probability of extraterrestrial intelligence is equal to the probability of extraterrestrial life, multiplied by the probability of life's evolving into mind. Knowing only of one form of life, which happens to be intelligent, we obviously lack the information for such a computation. All we have to go by are "guestimates," based on available knowledge, plausible surmise, and critical assessment, but unavoidably imprecise and exposed to personal bias. It was argued in Chapter 12 that the emergence of human intelligence by vertical evolution, although dependent on a large number of chance occurrences, was nevertheless subject to such stringent constraints as to make its probability much higher than is generally maintained. Even if this view is correct, there remains the question of how many life-bearing planets would provide the conditions allowing this evolution to take place. In a recent book,[3] two American scientists, geologist Peter Ward and astronomer Donald Brownlee, have defended the view that the number of conditions that had to be met simultaneously for higher animals, let alone humans, to arise on our planet is so high as to make it very unlikely that such an event could ever take place elsewhere. Their conclusion, it should be noted, concerns only animal life. "We believe," they write, "that life in the form of microbes or their equivalents is very common in the universe, perhaps more common than even Drake and Sagan envisioned."[4]

In principle, intelligence should be easier to detect than mere life, especially if it is manifested by the kind of technological civilization humankind has developed. Imagine an alien scanning our part of the world. The creature would have no difficulty recognizing strange signs on our blue planet, glowing at night with myriad artificial lights and, especially, ceaselessly throbbing with countless electromagnetic waves covering a wide gamut of wavelengths. At least there would be no difficulty provided the alien were close enough. From a faraway planet or spaceship, only the glare of the Sun would be detectable, totally obliterating the evidence of our existence. Should we wish to advertise our presence "cosmos-wide," we could not just rely on letting ourselves be discovered. We would have to send a message beamed out on a carefully selected wavelength and framed in such a way as to alert any observer's attention to the fact that something unusual is going on. This is what Frank Drake and Carl Sagan actually tried to do in the early 1970s, creating a strong protest on the part of the British astronomer Martin Ryle, who considered it reckless to thus betray our existence to hostile aliens who might use the information to launch an attack on us. Realizing that many years, if not centuries or millennia, might be required for their message to be received and answered, the defenders of the SETI project also bet on the chance that some alien civilization might want to get in touch with us in the same way.

What made these projects feasible was the development of radio-astronomy, a discipline that explores the skies by collecting invisible radio waves instead of light waves. One of the pioneers of this new technology was Martin Ryle, the scientist whose hostility to cosmic radio signalling was alluded to earlier. In Chapter 3, mention was already made of how the spectral analysis of incoming radiation, mostly in the centimetric wavelength region, has allowed the detection of a number of organic molecules and radicals in outer space. This type of radiation is also used for other purposes. It has, for example, allowed the discovery of pulsars (pulsating stars) and distant galaxies, whose light is too faint to be detectable with light telescopes.

Not that the signals are easily detectable; their power is so weak that only supersensitive instruments can record them. In a recent book, the British astrophysicist Martin Rees[5] recounts that visitors to Ryle's laboratory in Cambridge, England, were invited to pick up a tiny slip of paper on which was written: "In picking this up you have expended more energy than has been received by all the world's radio telescopes since they were built." In the words of Drake,[6] "all the energy collected in the history of radio astronomy barely equals the energy released when a few snowflakes fall on the ground." And he adds: "that's the energy released when they hit the ground, mind you; the energy lost as they melt is much greater."

Only the magic of modern amplifiers, combined with huge receiving surfaces, has made the detection of such weak signals possible. The most sophisticated such facility exists in Arecibo, in the northern hills of Puerto Rico. It boasts a shiny aluminum, bowl-shaped reflecting dish, 300 meters wide, with a collecting area of 80,000 square meters, capable of covering millions of channels at the same time. A new facility, already in an advanced stage of planning, will group 500 to 1,000 separate, small dishes over an area of one hectare (10,000 square meters) at some site in California. This "One Hectare Telescope" (1hT), which should be much more effective and cheaper than the Arecibo telescope, is expected to serve as a prototype for the "Square Kilometer Array" (SKA), of 100 times larger surface area.

In spite of all this technical wizardry, the SETI project remains an enormous gamble, as it does not just depend on the probability of extraterrestrial intelligence; it requires several other conditions. First, if it takes extraterrestrial life as long as it has taken terrestrial life to evolve intelligent beings, i.e., almost four billion years, all biospheres younger than that age at the time the signal is to be sent to be receivable on Earth today are excluded, as they cannot yet have reached the intelligent stage. A second condition is that the extraterrestrials should have attained a degree of technical development at least as advanced as ours and, in addition, should actually want to communicate with aliens such as us. That they have not yet done so has been quoted by a humorist as the best proof of the existence of intelligent extraterrestrials. A final important factor is the likely duration of an extraterrestrial civilization, which limits the time during which signals can be sent. Should they last only a few million years—the time allowed our civilization by some futurologists—the window of opportunity for our receiving a signal becomes very small. In spite of these uncertainties, the SETI project is still vigorously pursued, no longer supported by NASA but by private donations.

Colonizing Space

Scientific vocabulary has recently been enriched with a new word, "terraforming," which refers to the transformation of a planet in such a way that it becomes habitable. With Mars, for example, a first step would consist of warming the planet with "supergreenhouse gases," so that the carbon dioxide polar caps are sublimated into atmospheric carbon dioxide, which would subsequently help maintain a mild climate by its own greenhouse effect. Plants would then be introduced to generate the oxygen that, after an estimated 100,000 years, would allow our descendants to settle on the red planet.

This may sound like "superscience fiction." But the prospect of humankind progressively colonizing space has been entertained by a number of scientists and has even been taken seriously enough to alert "space ecologists" concerned about the protection of planetary environments. This preoccupation became a practical problem when the project to send a man to the moon began to take shape. Elaborate precautions were taken to minimize the danger of contaminating the moon with Earth germs as well as the reverse risk of bringing moon germs back to Earth. The exploration of Mars, which, unlike the moon is viewed as a possible abode of life, is raising even greater worries. In particular, if manned missions should land on Mars and spend some time there, contamination of the planet will be almost unavoidable, since astronauts can hardly be made germ-free. Fortunately, local conditions are such that the risk of a lasting implantation of human-carried Earth germs seems remote.

Given that we can imagine colonizing space, the possibility exists that extraterrestrials may have the same idea. Perhaps, if their degree of technological development is greater than ours, they could already have started implementing it. The invasion of Earth by extraterrestrials, which, starting with H. G. Wells's *War of the Worlds*, has inspired many works of fiction, is not pure fantasy; it is a possibility. The fact that it has not yet happened has even been used as an argument against the existence of extraterrestrial intelligence (the opposite argument has also been made in jest, see above). It is said that the Italian-American physicist Enrico Fermi, one of the prime builders of the atomic bomb, who was a firm believer in extraterrestrial intelligence, used to go around asking: "If they exist, why are they not here already?" To this question, often referred to as the "Fermi paradox," the Hungarian-born American physicist Leo Szilard, a colleague of Fermi's in the Manhattan project, famous for his wit, allegedly answered: "They are among us, but they call themselves Hungarians."

What to Szilard was a joke and to science fiction writers and movie makers has proved an inexhaustible source of imaginary drama, has been perceived as a true and often frightening reality by millions of people, ever since a pilot, Kenneth Arnold, flying his own plane near Mount Rainier on 24 June 1947, "saw" nine "flying saucers" cruising in his vicinity. The news made a sensation and was soon followed by other sightings of UFOs (Unidentified Flying Objects), some seen landing in a blaze of light and disgorging strange occupants. The buildup became so intense as to prompt the United States Air Force to commission an in-depth study of the question by a distinguished physicist, Edward Condon. As summarized by the American mathematician and professional "debunker" Martin Gardner, the 1,000-page "Condon report" concluded that "there are no UFOs that can't be explained as hoaxes, hallucinations, or honest misidentifications of such natural objects as meteors, Venus, huge balloons, conventional aircraft, reentering satellites, and atmospheric illusions."[7] Needless to say, confirmed "ufologists" and their believers are not convinced.

18. How About God in All That?

T HIS QUESTION WAS REPORTEDLY ASKED BY EMPEROR NAPOLEON the First of the physicist Pierre-Simon de Laplace, who had just explained to him the strictly deterministic principles of his *Mécanique céleste*. "Your Majesty," the famous French scientist is said to have replied, "I have no need for that hypothesis." Often denounced for its superbly disdainful assertiveness, this answer does, in fact, sum up the scientific attitude. Scientific inquiry rests on the notion that all manifestations in the universe are explainable in natural terms, without supernatural intervention. Strictly speaking, this notion is not an *a priori* philosophical stand or profession of belief. It is a *postulate*, a working hypothesis that we should be ready to abandon if faced with facts that defy every attempt at rational explanation. Many scientists, however, do not bother to make this distinction, tacitly extrapolating from hypothesis to affirmation. They are perfectly happy with the explanations provided by science. Like Laplace, they have no need for the "God hypothesis" and equate the scientific attitude with agnosticism, if not outright atheism.

The religions tend to make the same amalgamation, but in the recip-rocally negative sense. Science is godless and, therefore, is to be treated with deep suspicion. This sentiment reaches militant antagonism in the case of the more fundamentalist creeds. Other circles are more open-minded but remain profoundly distrustful. Remember, a bare 50 years ago, the Catholic Church still forbade publication of attempts by the French Jesuit Pierre Teilhard de Chardin to reconcile biological evolution with the teachings of the Church. It is only in October 1996 that the pope solemnly admitted that evolution "is no longer a hypothesis."

THE IMPOSSIBLE DIALOGUE

The traditional conflict between science and religion is understandable, considering that the two rest on entirely different premises. Science is based on observation and experiment, guided by *reason*. Religion is con-structed on a set of *beliefs*, taken by the Bible-inspired religions to be divinely revealed, with, in the Catholic Church, the additional guarantee of infallibility for the guardians of the faith. The two intellectual attitudes are so utterly irreconcilable that they can achieve peaceful coexistence only by ignoring each other. This, by and large, is what has happened. Scientists mostly do without religion. The reverse is obviously not true, as the practical applications of science are everywhere and cannot possibly be ignored by religion, were it only because of the many ethical problems they raise. But, through some strange dichotomy, religion rarely addresses the knowledge behind the applications, alleging incompetence. Or it does so only, as with the Big Bang, when it finds in the discoveries of science some apparent support for its beliefs.[1]

In the last few years, there has been something of a *rapprochement* be-tween science and religion. Part of it is due to a vocal minority of scientists, especially in the life sciences, who argue that science does not explain everything and that there must be "something else." I have mentioned some of these efforts in Chapters 3 and 12. A few other scientists, mostly physicists, claim to have discovered in their discipline the foundations for a revived "natural theology" and accordingly justify their allegiance to some religious system. Historians scrutinize the beliefs of Darwin and Einstein. On the religious side, an increasing number of theologians have become convinced that the discoveries of science can no longer be disregarded and must be faced head-on by the religions if these religions are to survive.

Thanks, notably, to the support of wealthy organizations, such as the Templeton Foundation, meetings bringing together scientists, philosophers, and theologians have multiplied.[2] Including the word "God" in the title of a science book is almost sure to make it a bestseller.

The present chapter may seem to be no more than a concession to fashion. It is not. Its topic has always preoccupied me. But it is true that the climate has changed. In the past, I have always chosen, for a variety of personal reasons, to confine myself to the scientific field, sticking to the objective exposition of the state of knowledge in the domains with which I am familiar and leaving readers free to draw their own philosophical or religious conclusions. I feel that this attitude is no longer justified today. I owe it to myself to face the implications of what I know and to express myself on the subject.

I mentioned above the basic incompatibility between science and religion. As long as this incompatibility persists, the dialogue between the two will remain impossible. For a dialogue, a common language is needed. This does not exist. And yet, such a dialogue has become more necessary than ever before. Science is in the process of transforming the world by its applications. And, especially, it is upsetting all our ideas about the nature of things. Religions, on the other hand, go on influencing human behavior in an extraordinary fashion, pervading all levels of society. It is urgent for the two to speak with each other.

It is tempting to say that this dialogue will be possible only by compromise. Let each add some water to their wine, and there will be understanding. Unfortunately, we are not dealing with a political or ideological conflict, but with respect for truth. On what has been convincingly demonstrated, science can make no concession. If there is conflict between what science *knows* and what religion *believes*, the latter must give in.

This conflict has become particularly acute in the domain of life, in which a widening gap separates the discoveries of science from a number of notions contained, explicitly or implicitly, in the religious message. The time has come to compare knowledge and beliefs in order to find out in what measure the latter need to be revised in the light of the former. Because of my personal background, my comments on religion will be drawn largely from what I know of the Catholic religion. Readers more familiar with other religions should have no trouble making the appropriate transposition.

LIFE'S MESSAGES

Throughout this book, I have attempted to summarize the main advances of the last decades in our understanding of the nature and history of life. What now of the implications of these advances for religious beliefs?

THE NATURE OF LIFE

As I hope to have shown, the proof that life is a *natural* manifestation of matter that takes place without the help of any sort of vital principle is overwhelming. Adopted as a matter of course by the great majority of contemporary biologists, this mechanistic vision of life has yet to become commonly accepted knowledge among the general public. Contradicting, as it does, the deeply ingrained notion of "animated matter" that a millennia-old tradition, perpetuated by poetic language and by religious vocabulary, has anchored in the human imagination in a form that still permeates much current thought and discourse.

It is true that vitalism is not necessarily included in a religious outlook, which may well, for example, accept the Cartesian notion of animal-machine. Nevertheless, the feeling that the functioning of life involves something other than purely physical-chemical processes is often associated with religious belief. This feeling sometimes even transpires, at least tacitly, in the declarations of scientists. When reading the philosophical musings of certain physicists and cosmologists, I am struck by their vision of life, depicted as some strange, extraneous phenomenon, generator itself of something even stranger, conscious thought, all of this somehow pinned on an unfeeling background of swirling galaxies without truly belonging to it. Such a view is wrong, a point that cannot be emphasized strongly enough. Life is part of the universe; it is a normal manifestation of matter and obeys the laws of matter; it is explainable in terms of those laws and can, accordingly, be manipulated by agents subject to them.

THE UNITY OF LIFE

Another notion that may be considered as established with a high degree of certainty is the kinship among all living beings known on Earth. All, including humans, are descendants from a *single ancestral form*, from which they have inherited all their shared basic properties. Although few probably would deny the deep similarities that unite all living beings, the historical origin of these similarities is far from being unanimously accepted. As I shall mention later, the reality and mechanisms of biological evolution continue to feed discussions and controversies in religious circles.

THE ORIGIN OF LIFE

In this book, I have defended the thesis, accepted by the vast majority of scientists, that life *arose naturally*, by the sole enactment of physical and chemical laws. This thesis runs counter to the belief, supported with more or less vigor by many religious bodies, that special, divine intervention was needed to "breathe life into matter." To strict creationists, such an intervention leaves no doubt. For the more liberal religions of the Judeo-Christian tradition, including the Catholic Church, it is not an article of faith. Neither, however, is such an intervention explicitly expurgated from current discourse, which frequently confounds animism and religious belief in a vague mixture very few take the trouble to clarify. In my experience the immediate reaction of lay audiences, even highly educated ones, when told that life arose naturally, is often one of disbelief, if not distrust. The notion is seen as a dangerously materialistic deviation. Or else it is dismissed and put away with those other incomprehensible peculiarities that keep scientists busy but do not concern the common run of people.

One must admit that the spontaneous generation of a living organism from nonliving matter has never been observed in nature or produced experimentally. There is no direct proof, therefore, that such a phenomenon occurred or, even, is possible. The notion that life arose naturally just happens to fit with all we know of the nature of life and it is supported by a variety of observations and experimental data; it is an almost obligatory corollary of the abandonment of vitalism and the only working hypothesis capable of guiding research in a fruitful way. On the other hand, the objections put forward against this notion by the defenders of intelligent design do not stand up to objective scientific analysis, as I hope to have shown. Yet, I feel it would be a mistake to make this point into an issue between science and religion. The science of today cannot prove wrong those who wish to attribute the origin of life to divine intervention. Science can only point out, as I have done in Chapter 3, that such an intervention appears unnecessary, as well as unlikely, in the light of present knowledge.

THE EVOLUTION OF LIFE

Evolution is a *fact* that is now established beyond reasonable doubt. So is its main mechanism by *natural selection* acting on accidental genetic modifications *devoid of intentionality*. The findings of molecular biology can leave no doubt in this respect. It is true that experts still find much to disagree on within this general framework. Sometimes mistakenly brandished by the adversaries of neo-Darwinian theory as evidence against this theory, such disputes concern details of the theory, not its substance. Exceptions are those few scientists who defend a finalist view of evolution and claim that some key steps in the history of life could not have taken place without the help of a guiding principle, which some do not hesitate to identify with the hand of God.

Today, the fact of evolution is accepted by most religious bodies, including, as we have seen, the Catholic Church; it is negated only by those who, like the strict creationists, willfully blind themselves. As to the mechanism of evolution, religions rarely take a clear stand for or against the modern theory. But there is no denying that they look with considerable sympathy upon the concept of intelligent design and tend greatly to exaggerate the importance and significance of the movement supporting this concept. I know this from personal experience, through my attendance at recent meetings organized to promote the science-religion dialogue.

THE ADVENT OF HUMANKIND

When it comes to the origin of humankind, the rift between science and religion becomes much wider, to the point of being virtually unbridgeable without a major sacrifice on the part of one or the other. It is one thing to accept the reality of evolution, but quite another to reconcile this reality with the concept of the human person endowed with an immortal soul and created in the image of God, a keystone of the major monotheistic religions. This has been clearly understood by the fundamentalists, for whom every word of the Bible is to be taken literally. In a certain sense, they are the only ones holding a coherent discourse. They correctly appreciate that, once a concession is made, the way is open to further weakening of the doctrine. Thus, once one accepts the notion that humanity arose through a natural evolutionary process, one is faced with a series of problems: At which stage of evolution was a hominid converted into an authentic human being, endowed with an immortal soul? Was

Lucy human? What about Neanderthal man? Where lies the discontinuity, if there is one, in the continuum of evolution?

A similar question applies to embryological development in relation to practices such as voluntary interruption of pregnancy or research on human embryonic material. At what moment in the development of a fertilized egg does an embryo become a true human being? Note that this issue is not just a problem for theologians; it also concerns lawmakers, who are asked, for example, whether it is permissible to deliberately destroy human embryos or use them for experimental purposes or at what stage of development an abortion becomes an infanticide. Here, again, a discontinuity has to be set artificially in what is essentially a continuous process.

The Catholic Church has not failed to recognize these difficulties. In his historic speech of 26 October 1996 accepting evolution, John-Paul II took care immediately to add that "the magisterium of the Church is directly interested in the question of evolution because it touches on the concept of man, of whom Revelation teaches us that he was created in the image and resemblance of God." He points out further that human persons owe their dignity to the fact that they possess a spiritual soul and quotes Pius XII, who "emphasized that essential point: whereas the human body owes its origin to the living matter that exists before it, the spiritual soul is immediately created by God."[3]

In the same message, the pope briefly addresses the problem of what he calls "the passage to the spiritual." He acknowledges the difficulty of reconciling the "ontological discontinuty" of humankind with the "physical continuity" of evolution, but he eludes the question by invoking two different methods of knowing: experimental science for the latter, philosophy and theology for the former. To science, as we have seen in Chapter 13, the discontinuity is an artifact produced by the absence of the many intermediary forms that have landmarked what is essentially a continuous process. It is the juxtaposition of beginning and end that creates the impression of a jump.

BRAIN AND MIND

The discontinuity problem also arises with respect to mental faculties. Religious doctrines are deeply permeated with dualism. Although not necessarily an article of faith, attribution of mental faculties to the soul is implicit in the religious discourse. For the vast majority of believers, it is the responsible self that survives after death, with the weight of its sins and the benefit of its merits. In the message already cited, John-Paul expresses himself clearly on the topic, stating that "theories of evolution that, as a function of the philosophies that inspire them, consider spirit as emerging from the forces of matter or as a mere epiphenomenon of this matter are incompatible with the truth of man. They are, moreover, incapable of providing a foundation for the dignity of the person."

We have seen (Chapters 12 and 13) that the ideas of scientists on this subject are very different. Even a Christian philosopher such as Jean Guitton, who was known as a friend of popes, has not hesitated to decide in favor of a monistic concept denying any opposition between matter and spirit (see citation p. 220, in Chapter 14). As far as I know, the Church has not paid any attention to Guitton's declaration, whether to condemn it or to incorporate it into its teaching, perhaps applying the principle that what disturbs is better ignored.

LIFE AFTER DEATH

Of all the beliefs propagated by religion, that of survival after death is probably the most difficult to reconcile with scientific data. But, at the same time, it is the belief humans most ardently cling to. For nothing is more difficult to accept than the definitive character of death. A friend, left disconsolate by the loss of his wife a few years ago, asks me: "Do you believe I shall ever see her again?" I don't have the courage to tell him what I believe. I answer: "I don't know," which, after all, is the truth. A very dear woman friend, a widow for more than 20 years, does not even ask the question. When she looks at the sky, she "sees" him, beyond the clouds. Why should I tell her that there is no room in the cosmologists' sky for a place where the resurrected could indefinitely pursue their terrestrial existence, with all its joys, but without its sufferings and vicissitudes? She knows it, but she puts her faith above her knowledge, trusting in the infinite compassion of a God who can do anything.

Ever since humans became conscious of their own mortality—they are believed to be the only animals to possess this knowledge—they have refused to accept its irrevocable nature. Witness, across human cultures, the graves, funeral rites, beliefs in one or the other form of survival, be it resurrection, reincarnation, metempsychosis, or some other existence, whether happy or painful, in the "abode of the dead." Not so long ago, the Catholic Church still forbade the incineration of cadavers, a practice

suspected in the past of possibly interfering with the chances of resurrection of the dead person. I still remember a book I read when I was a
student, in which the author triumphantly acclaimed the discovery of
isotopes as possibly explaining how everyone might recover his or her
own atoms when rising from the grave. When I shared this jewel with
one of my physics professors, expecting him to join me in a gigantic
guffaw, he responded in all seriousness: "Why not?"

EXTRATERRESTRIALS

The possibility, which we have seen is seriously entertained by a number
of scientists, that intelligent beings may exist or arise elsewhere in the
universe is another subject of concern for theologians. Are such beings
sullied by Original Sin and thus in need of salvation by a Redeemer? Or
have only terrestrials known the Fall? In discussions of such topics that
I have attended, this question seemed particularly to worry clerics, to the
point that most of them prefer to avoid it by holding fast, until proven
otherwise, to the hypothesis of the uniqueness of the human species.
That humans may be only an intermediary link in evolution, rather than
its outcome, also disturbs. The very nature of Original Sin raises serious
problems as well. Once the biblical story is accepted as mythical, how
can the need for a Redeemer still be justified?

Attempting to answer those questions may prove an embarrassing
job. Witness the following declaration: "At the very beginning human
beings did something bad. They revolted against the God who had
made them." Astonishingly, this sentence is not excerpted from some
Sunday-school manual but from a Keynote Paper, entitled "Science and
Religious Belief," with which the Jesuit George Coyne, a renowned
American astronomer who heads the Vatican Observatory in Castelgan-
dolfo, opened an international symposium in June 1998.[4] In this allo-
cution, the distinguished priest-scientist addresses, mostly in the form of
interrogations to which he supplies no answers, two of the problems
raised earlier: the uniqueness of humankind in the context of its evolutionary origin and the existence of extraterrestrials in relation to Original Sin. The sentence quoted introduces this second topic. What follows
is no more enlightening. No mention is made of how the alleged revolt
can possibly fit within the early history of humankind as retraced by
paleoanthropology.

CONCLUSION

The facts speak for themselves: *several of the teachings of religion are incompatible with the discoveries of modern biology.* Faulting science and rendering it responsible for the contradictions, as some fundamentalists would have it, would negate the value of the scientific approach, with all its careful safeguards and rigorous precautions. It would also ignore the whole edifice of practical applications built upon the knowledge achieved by this approach. No intellectually honest person can accept that. Truth cannot be evaded. Surely, the mistakes must be in the religious accounts. This is hardly surprising in view of the historical context within which these accounts were first conceived, at times when myths prevailed and animist explanations of natural phenomena were accepted as a matter of course, unquestioned by even the most enlightened thinkers. The question is: What should be done about it?

AN AGONIZING BUT INESCAPABLE REAPPRAISAL

The answer to the question asked above is clear. Religions must revise their scripts and bring them in line with modern science. This, however, is more easily said than done. A number of factors conspire to render such revision very difficult.

THE ROAD TO CHANGE IS STREWN WITH OBSTACLES

Churches are large, rigid bodies, often highly organized in hierarchical structures dominated by powerful authorities. Anybody with some inside knowledge of the Vatican can testify to this. Such structures have an enormous resistance to change, all the more because they are not, in spite of the lofty tone of their discourse, immune to the personal rivalries, vested interests, and other corrupting influences that inevitably accompany power. Even the more "democratic" religious bodies rarely allow the kind of unfettered discussion that is customary among scientists. Not for nothing has the phrase "Rock of Ages" been chosen as an emblem of religious immutability.

Another factor to be taken into account is the immense power of faith, a sentiment capable of "lifting mountains," of inspiring the noblest of sacrifices, up to martyrdom, as well as the most cruel persecutions and ruthless wars. By definition, faith does not rest on rational grounds, even though attempts may be made to rationalize it *a posteriori*. It is based on the blind acceptance of authority, which itself claims to be enlightened by "Revelation." Not without reason is faith described as "a gift from God," not to be questioned. Yet, it often proves more powerful than reason. There is probably an evolutionary explanation for this. Most likely, those human groups that believed in something were better able to survive and propagate their own than those that did not, no matter whether what they believed in was true or not.

Faith would not have its power without human credulity, probably retained by natural selection for the same reason that there were more advantages in believing in something than there were disadvantages in being fooled. Whatever the explanation, there can be no doubt that the ability to believe is stronger than the ability to listen to reason. The success of esoteric sects, horoscopes, crystal gazing, card reading, faith healing, and other exploitations of gullibility, even in highly sophisticated societies, clearly illustrates the strong human propensity to believe "without asking the reason why." It is striking, but probably inevitable, that groups that propose a belief system generally claim to be the sole holders of the truth, branding all other competing creeds as heresy or superstition. Such "truth monopoly" is a major stumbling block in all attempts at ecumenism. Strangely, few of the adherents to a system seem to be disturbed by such illogical behavior, so great is their confidence in their own version of the truth, or Truth, as many religions spell this word.

To these factors must be added the intellectual stratification that exists in many churches, a *de facto* response to the educational stratification of their members. Religion is addressed to everybody and must, perforce, adapt its language to the degree of literacy of its faithful. What is taught in a Catholic university is not necessarily promulgated from a parish pulpit. Theologians and pastors rarely communicate with each other. It is not uncommon to hear that certain sensitive issues are better kept to the learned discussions of philosophers and theologians, so as not to shake the confidence of the naïve faithful.

Finally, a significant facet of this problem lies in the important social functions fulfilled by religions. In many parts of the world, churches play a major role in education, health-care delivery, and other forms of welfare. Churches also offer support and solace in many personal or family ordeals, such as bereavement, sickness, disability, estrangement, professional failure, or financial misfortune. Conversely, the association of churches with joyful events, such as births and marriages, is also valuable.

Even more important, churches generally provide the main ethical guidelines that rule the conduct of their members and often spill over into legislations. One may well ask what will happen to these good works and beneficial activities if their ideological and sentimental underpinnings are sapped. For scientists, this question raises a major case of conscience.

MUST SCIENCE SPEAK?

Faced with the persistence of myths and their consoling virtues, scientists must ask themselves whether they have the right—or is it a duty?—to disabuse those whose beliefs are incompatible with what they, the scientists, see as undeniably established. I have long hesitated to do so, out of respect for the opinions of others, out of loyalty to the institution to which I belonged, and also out of scientific caution. In science, we rarely feel sure enough to affirm. But there are limits to such scruples. As already mentioned, I have decided to speak more openly on these matters, prompted by the importance of the issues and encouraged by the present trend favoring the dialogue between science and religion.

Unfortunately, the message, as we have seen, is profoundly disturbing. It shatters the age-old vision that places our human species at the center of a world created for its sole benefit. It questions a tightly interwoven fabric of relationships, behaviors, and beliefs that unites and consolidates vast human groups, with the help, no doubt, of natural selection, which has retained the underlying dispositions for their strictly utilitarian value, regardless of their correspondence with reality. To tear this fabric is an agonizing act few are prepared to attempt. On the other hand, to conceal facts for the sole reason that they might conflict with treasured beliefs is ethically indefensible. More to the point, it is an insult to one's fellow human beings, who thereby are implicitly treated like children too immature to face the truth.

WE MUST PREPARE THE FUTURE

As the readers will have noticed, I take these matters seriously. But I hardly expect Earth-shaking results. Like the many efforts of a similar kind by other scientists, my contribution is likely to reach mostly those who are already convinced. But we can only try. Given enough time and persistence, the message will eventually be heard. In the meantime, we can start thinking about what the message of the future should be. It is not enough to point out mistakes. Something positive must be proposed that can eventually replace the myths propagated by religions, while trying not to destroy the many beneficial structures religions have built on the myths. Is there anything in the new vista opened by science that can inspire such a worldview? Many scientists have asked this question, coming up with answers that vary from the bleak and despairing to the optimistic and hopeful.

WHAT DOES IT ALL MEAN?

In his 1977 account of *The First Three Minutes* in the history of the universe, the Nobel prize-winning American physicist Steven Weinberg tells of contemplating Earth from an airplane and reflecting: "The more the universe seems comprehensible, the more it also seems pointless."[5] Another distinguished American physicist, British-born Freeman Dyson, writes in his 1979 *Disturbing the Universe*: "The more I examine the universe and study the details of its architecture, the more evidence I find that the universe in some sense must have known that we were coming."[6]

Thus, two eminent scientists, equally knowledgeable about our present understanding of the ultimate properties of matter, almost at the same time voice radically opposed opinions on the meaning of what they know.[7] Biologists also have their disagreements, even when they agree on the facts themselves and their interpretation. This is not surprising. As soon as we move from facts to their significance, we leave the domain of science. We no longer face problems that can, at least in theory, be solved by the objective and rational examination of available data. We enter the dim area of the subjective: creeds, biases, feelings, and other inner experiences, strongly colored, to be sure, by the practice of science, but also influenced by many personal factors. I can hardly hope to avoid such biases, though they be unconscious. Here, for what they are worth, are a few reflections that come to my mind as I contemplate, as objectively and dispassionately as I can, the present state of our knowledge of life and of its place in the universe.

CONTINGENCY IS A RED HERRING
One of the most pervasive themes of modern thought is the overwhelming debt we humans owe to chance. A committed and persuasive defender of this view, the American paleontologist and popular science writer Stephen Jay Gould, whose death, at the early age of 60, was an-

nounced as this book was going to press, expressed it in the following words: "Biology's most profound insight into human nature, status, and potential lies in the simple phrase, the embodiment of contingency."[8] Presented as incontrovertibly enforced by the findings of biology, this notion has fed a number of philosophical considerations that have in common a belittling of the human condition and a denial of its significance.

As early as 1970, French biologist Jacques Monod concluded his best-seller *Chance and Necessity* with the comment, much in line with the existentialist ideology of the absurd greatly in favor in the France of his time: "Man knows at last that he is alone in the Universe's unfeeling immensity, out of which he emerged only by chance."[9] In the book cited above, Weinberg likewise speaks of human life as "just a more-or-less farcical outcome of a chain of accidents reaching back to the first three minutes" and of Earth as "just a tiny part of an overwhelmingly hostile universe."[10]

More radical—and pernicious—than these melancholy additions by scientists to a long poetic tradition of bewailing the fragility of human-kind is the human-bashing movement alluded to in Chapters 8 and 12. This movement is now being propagated as an unavoidable outcome, however unpalatable we may find it, of the discoveries of science and has come to be viewed as politically correct in a number of influential circles of the intelligentsia.

A major objective of this book has been to expose the fallacy of this "gospel of contingency," which is being preached in the name of science. The alleged scientific premises of this doctrine, as I have tried to show, are incorrect. Not, as some would have it, because there is "something else" shaping the direction of evolution, but because the natural con-straints within which chance operates are such that evolution in the di-rection of increasing complexity was virtually bound to take place, if given the opportunity. Chance does not exclude inevitability.

THE HEART OF THE MATTER
The argument from contingency not only rests on dubious premises; it is debatable in itself. Irrespective of probability estimates, the mere fact of our existence remains, in my opinion, supremely significant. Whatever sentimental or ideological reasons one may have for lamenting or deni-grating the human condition, calling on science to bolster them is

unwarranted. Contrary to the frequently asserted view, the probability—
or improbability—of life and mind is philosophically irrelevant. Whether
life and mind are commonplace, exceptional, or even unique in the universe is immaterial with respect to the central fact that life and mind
exist. There lies the heart of the matter. Monod's saying—"the Universe
was not pregnant with life, nor the biosphere with man"[11]—is logically
flawed. It is self-evident that the universe was pregnant with life, and
the biosphere with man. Otherwise, we would not be here. Or else our
presence can be explained only by a miracle or, rather, two miracles—
two successive births without pregnancy, in violation of the laws of the
universe—which is certainly not what Monod had in mind.

An old English aphorism says that "you can't make a silk purse out
of a sow's ear." The French have a more poetic—and Gallic—way of
putting this: "The most beautiful girl in the world can give no more than
what she has." The converse of this saying is that whatever she gives she
must have had in the first place. The universe has given life and mind.
Consequently, it must have had them, potentially, ever since the Big
Bang. What this fact implies has been the object of much discussion in
recent years.

A UNIVERSE MADE TO ORDER?
A key element in this discussion lies in what is implied, in terms of "fine
tuning" of the universe, by the "pregnancies" of Monod's saying. Detailed
calculations by physicists have shown that if any of the four fundamental
atomic constants, defining the strong, weak, electromagnetic, and gravitational interactions, had values even slightly different from what they
are, our universe could not have produced the material conditions needed
for life to arise, subsist, and evolve. A similar case has been made for a
number of cosmological properties, such as the size, curvature, total mass,
and isotropism of the universe as a whole, as well as for the value of a
constant, known as the cosmological constant, which has a critical effect
on the mode of expansion of the universe. These considerations have
been embodied under the name of *anthropic principle*, a term derived from
the Greek *anthrôpos* (human being), not to be confused with the entropic
principle, which is the second law of thermodynamics.

In the view of many advocates of the anthropic principle, the extraordinary set of coincidences that has conspired to make life and mind
possible indicates that, for some deeply hidden reason, the universe *had
to be* the way it is. Some even go one step further and see in it proof

that the universe has been *designed* to be so. There is another explanation, however, which is now being considered by physicists.

According to this new theory, our universe is just one among an enormous population of universes. A tiny, insignificant component of an infinitely large "multiverse," the felicitous term coined by the British astronomer Martin Rees, one of the main proponents of the theory. The chance product, like all the other universes, of a random fluctuation in something described as "chaotic vacuum." But, contrary to the vast majority of those other universes, it would happen to be, through the mere chance operation of the law of large numbers, endowed with cosmological properties such that it can produce life and mind, and thus be knowable by beings of its own making. It could even be the outcome, by some kind of cosmic natural selection process, of a huge evolutionary game, as proposed by the American cosmologist Lee Smolin.[12] Such views tend, implicitly and, sometimes, explicitly, to trivialize our universe by diluting it with myriad others that do not share its unique properties. As told by Rees, "once we accept this, the seemingly 'designed' or 'fine-tuned' features of our universe need occasion no surprise."[13]

Weinberg, the American physicist already mentioned earlier, adopts the same stand against a fine-tuned universe. "The expanding cloud of galaxies," he writes, "that we call the Big Bang may be just one fragment of a much larger universe in which big bangs go off all the time, each one with different values for the fundamental constants." And he concludes, echoing Rees: "In any such picture, in which the universe contains many parts with different values for what we usually call the constants of nature, there would be no difficulty in understanding why these 'constants' take values favorable to intelligent life."[14]

Perhaps, some day, new advances in cosmology and in theoretical physics will allow the multiverse theory to be tested. Even if the theory should turn out to be correct, the deduction drawn from it by Rees and Weinberg strikes me as what is called in French "drowning the fish." Whether you use all the water in the oceans to drown the animal, it will still be there, affirming its presence. However many universes one postulates, ours can never be rendered insignificant by the magnitude of this number. Whether the combination of constants that allowed life and mind is unique or only one in a very large number of combinations most of which lack this property, what to me appears as supremely significant is that a combination capable of giving rise to life and mind *should exist at all*. Life, mind, and the "superminds" that may arise, later or elsewhere, are such extraordinary manifestations that their emergence, by whatever mechanism, cannot but be a telling revelation of ultimate reality.

THE FALLACY OF ANTHROPOCENTRISM

Yes, there is something very special about our universe's being pregnant with life, which is itself pregnant with mind. But that does not make it a universe "made for us," as is implied by the anthropic principle. One of the dominant themes of this book has been the transience of humankind. Less than one million years ago, we were not around. A few million years from now, we may very well not be here any more. It is virtually certain that there won't be human beings, as we know them, a few hundred million years from now. On the other hand, it is quite possible—I tend to say probable—that other beings considerably more mentally advanced than we are will be present on Earth by then.

To beings with a life expectancy of the order of 100 years at best, such considerations are of no practical interest, except, possibly, in the framework of what we may do to shape the long-range future. But they are highly relevant with respect to the possible cosmic significance of humankind. The lesson I draw from them is that we are not the final outcome of evolution, but only an intermediary stage, perhaps even a blind alley, in an ongoing process that is likely to continue for at least 1.5 billion years and to lead to beings we are totally unable to imagine.

This humbling message is different from that propagated by the advocates of the "gospel of contingency". It does not reduce the human condition to a meaningless toy tossed by chance on the waves of uncertainty. It does not oppose the view that places humankind on top of the tree of life (for now) and sees the attainment of this position as a significant, perhaps even necessary, step in the unfolding of biological evolution. What the message stresses is the *relative* and transient character of that position.

The central fallacy is anthropocentrism, an understandable but undue extrapolation of humanism, as we have seen in Chapter 16. The notion that the whole world revolves around us is deeply ingrained in human nature and has colored human thought for millennia. That this notion may be false has only recently been brought to our attention and has yet to be integrated in our worldview. Anthropocentrism is also a keystone of many religions, especially those inspired by the Bible, which describes man as created *in the image of God*. No phrase could be more quintessentially—and arrogantly—anthropocentric than this excerpt from the Book of Genesis. According to this Book, man is, by divine right, master of the creation, accountable only to the creator. Theocentrism prevails over anthropocentrism. But even here, the *anthrôpos* component plays an overwhelming role.

THE MANY FACES OF GOD

It all started with gods, imaginary beings that were invented to explain all kinds of natural events and that could be invoked or needed to be placated, sometimes with horrible sacrifices, in order to bend their power in a desired direction. The phenomenon is universal. There is virtually no people in the world that has not—or, sometimes, still has—its collection of gods. Their numbers, names, and attributes vary greatly, but they have in common that they are in some way humanlike, however extravagant their behaviors and magic their powers. The whole of mythology is filled with their antics, which entertain us today but seem to have been taken seriously by their inventors, even such highly sophisticated peoples as the ancient Greeks and Romans.

The advent of monotheism was an enormous progress over polytheism. But personification remains a key attribute of the now unique God. The God of the Bible resembles in many ways the ruthless kings of the time. He is domineering, vengeful, jealous, merciless. He is even something of a macho. By creating Adam before Eve, he establishes sexual discrimination right from the very start, a disparity that has been conserved up to this day by all three major monotheistic religions.

The image of God has mellowed, notably with the advent of Christianity but the notion of an all-powerful Lord has been retained, with as major modification a greater emphasis on mercifulness. Today's God—at least according to the Judeo-Christian tradition—cares for His creatures. He follows their efforts toward goodness with loving solicitude and even witnesses their wrongdoings with compassion, ready to forgive if sincere contrition is expressed. He listens to prayers and sometimes, when He so chooses, uses His power over nature to grant them. On rare occasions, He actually manifests Himself, or allows one of His elected creatures to do so for Him, by way of an authentic miracle. Furthermore, this God, after long remaining hidden, has decided, for mysterious reasons of His own, to reveal His existence and dictate His laws at a particular stage in the history of humankind, roughly 3,500 years ago. Some

1,500 years later, so the Christian version goes, He sent His Son to Earth in human form, there to announce the good tidings and suffer the ultimate sacrifice exacted by Salvation.

To hundreds of millions of people—more than one billion for the Old Testament part—this account is a matter of deep and sacred faith. While I respect this sentiment and sympathize with it, having shared it in my youth, I cannot help questioning its verisimilitude. Based entirely on documents dating back to times when anthropomorphic mythology and animist explanations dominated human thought, the biblical account clearly reflects the very human imaginativeness and preoccupations of its authors. The God of the Bible is a person, exhibiting, in appropriately perfect form, all the qualities seen as desirable in a human ruler with, in addition, the possession of supernatural powers. Whether described as almighty Creator, stern Ruler, inflexible Judge, compassionate Shepherd, loving Father, or a mixture of all these, the God of the Bible remains rooted in the deceptive imagery of wishful anthropomorphism. The sentence from the Bible quoted earlier should be reversed. It is man who created God in his own image.

There are many, even among believers, who have come to reject as obviously mythical the anthropomorphic image of the biblical God but, nevertheless, describe themselves as *deists*. For them, the concept of God is mostly associated with that of the Creator, the great Architect, the divine Watchmaker. The images vary, but the content is the same: a Supreme Being who made the world and exists outside it. Such a view is held by a number of scientists. It is compatible with science in the measure that the postulated God, after flipping the universe into being, merely sits back and lets His creation unfold without interfering with its operations. Intelligent design is restricted to the Big Bang and is revealed in the remarkable set of coincidences uncovered by the anthropic principle. This explanation is intellectually satisfying to the extent that it offers an answer to the key metaphysical question raised by the existence and properties of the universe. But it still retains a flavor of anthropomorphism, were it only in its imagery.

Most scientists go one step further in their denial of a personal God. Many simply don't bother with the problem and declare themselves *agnostics* (from the Greek meaning that they don't know). Some explicitly deny the existence of any kind of God and even defend their *atheism* with militant zeal. Among them are Steven Weinberg, the American physicist quoted earlier, the British chemist Peter Atkins (in *The Crea-*

tion,[15]) and the British ethologist and evolutionist Richard Dawkins (in *The Blind Watchmaker,*[16] one of his bestsellers). So also is the Swedish immunologist of Hungarian origin George Klein, who has made an interesting comment about atheism. In a book significantly titled *The Atheist in the Holy City*, Klein refers to a letter from a friend calling him an agnostic, not an atheist, and quotes his answer: "I am not an agnostic. I am indeed an atheist. My attitude is not based on science, but rather on faith. . . . The absence of a creator, the nonexistence of God is my childhood faith, my adult belief, unshakable and holy."[17] Explicit disbelief, in other words, is a form of belief. Atheism is in some way a religion.

My own position I find difficult to define. I could take refuge in agnosticism, except that this position appears to me a cop-out, a comfortable way of evading the issue. If pressed, I refuse to describe myself as an atheist. Yet, I am unable to subscribe to the notion of an anthropomorphic God. In my view, we must "depersonalize" God, just as the new physics tells us we must "dematerialize" matter. To me, there is no other term in our language for the entity that will emerge in this way than "ultimate reality."

ULTIMATE REALITY

This term has cropped up on several occasions in this book. It applies most pertinently to the discoveries of science. Thanks to the extraordinary and continually growing power of the scientific approach, we are beginning to learn something of what lies behind entities such as the cosmos, matter, life, and mind. In this process, we have been forced to transcend appearances, in order to assimilate strange new concepts almost irreducible to our familiar world. This exploration is far from complete. But few will doubt that it is bringing us closer to the reality behind the appearances.

Only a few "deconstructionist" philosophers contest this point, claiming that scientific knowledge is as relative and socially loaded as any other attempt to understand reality, be it myth, conjecture, transcendental meditation, or, for that matter, philosophy. There are not many, however, even among philosophers, who are ready to accept such negativistic pronouncements. The validity of the scientific method as a self-questioning and self-correcting approach to the truth is almost universally recognized. Few maintain that what we have learned from modern physics, cosmology, and biology has not led us to a better understanding of the universe and of our place in it.

What about our other mental preoccupations, the aspiration after beauty, the sense of right and wrong, and the yearning for love, which, with the search for truth, seem to be universal constants of human nature? The existence of these preoccupations raises a question of fundamental importance, already addressed in Chapter 13. Are they no more than strictly utilitarian properties, unrelated to any sort of objective reality, that were retained by natural selection because individuals and groups who experienced them survived better and produced more progeny than those who did not? Or on the contrary, do these preoccupations reflect our perception of authentic facets of ultimate reality that are not accessible to the rational intellect but nevertheless exist in their own right?

This question goes back, in a different conceptual context, to the ancient Greeks and to the opposition between Plato and Aristotle. I see no scientific way of settling the issue. I can only confess to an intuitive bias—or is it an early imprinting?—in favor of the Platonic view. As I have attempted to convey, I see ultimate reality as expressing itself through several facets to which we each are more or less finely attuned depending on the particular structure of our cortical polyneuronal networks, as they have been wired by heredity, experience, and education. It is the rare privilege of a few—the great thinkers, scientists, poets, artists, mystics, and spiritual leaders—to possess particularly receptive networks and to be able to communicate their experiences to the rest of us. But even they remain hemmed in like the rest of us by the exiguity of the human cerebral cortex and are afforded only nebulous glimpses of this reality. Even with their help, we can discern no more than those glimpses, just enough to fill us with wonder, yearning, and a feeling of being part of something entirely beyond us, but meaningful. Our apprehension is keener than Lucy's but still woefully inadequate.

THE FUTURE OF RELIGIONS

This book is about knowledge, understanding, meaning, and belief. It was my intention to end on the same plane. Then came September 11 and, now, the increasingly bloody confrontation of Jews and Muslims in the Middle East, let alone the perennial slaughter of Christians by Christians in Northern Ireland and numerous examples of ideological strife all around the world.

It would be a facile oversimplification to blame all these conflicts on religion. Upholding the Faith is often but a pretext for conquest, plunder, and subjugation. Yet, without the promise of eternal heavenly delights, there would be fewer suicide bombers. Colonies would be implanted with less self-righteous zeal in erstwhile Judea and Samaria if they were not supported by ancient biblical claims. Not to forget the innumerable crimes committed under the banner of the Cross.

Now that their doctrinal underpinnings have been invalidated, should not the religions themselves be eliminated at the same time? This, certainly, is the opinion of the more aggressive atheists and agnostics among my scientific colleagues. I am not ready to follow them that far.

First, trying to eliminate religions would be a totally unrealistic undertaking. Remember the immense power of faith, the almost boundless credulity of even educated persons, the rock-like solidity of religious bodies. It would be fatuous to expect one more appeal to reason to suddenly reverse an overwhelming social trend, with roots going back to the earliest days of humankind. Anthropology tells us that the religious phenomenon is ancient and universal. Many prehistoric artifacts bear witness to what appear to have been myths and rituals of one sort or another, going back tens of millennia. No people is known that does not entertain some kind of belief, often associated with magic. The religious feeling is deeply embedded in our nature, probably carved into it by natural selection.

This, it may be argued, is true of other traits—aggression is an often-quoted example—that proved useful to our hunter-gatherer ancestors but are no longer adapted to modern societies. The fact that a trait has been retained by natural selection hardly justifies our not combating it if it has become undesirable. But is the religious feeling of that kind? That is far from certain. In my opinion, the religious feeling answers a real need and may even correspond to an authentic reality. It expresses our wonders, our desires, and our yearnings. We have to change the clothing in which we cloak the feeling, but we need not abolish or stifle it. We probably can't, in any case.

Erected in an upsurge of faith and fervor, our cathedrals are sometimes compared to empty vessels that have lost their *raison d'être*. This, in my view, is a mistake. We need churches, as we need laboratories, museums, theaters, and concert halls. It is good that men and women gather in churches, no longer to solicit favors from a human-made God, or even revere Him, but to contemplate and meditate. Some prefer solitude for this exercise. But to many, communion with others in an atmosphere of shared devoutness is a means to forget the mundane preoccupations of the day and merge with mystery.

Priests are still needed, as are thinkers, scientists, philosophers, poets, writers, musicians, sculptors, painters, and other artists, as well as performers of various kinds. But not priests who drape themselves in the mantle of authority and claim to hold the truth by a direct line with God, up to the ultimate presumption of granting themselves a certificate of infallibility, subsequently enforced as an article of faith. Even less do we need priests who exploit the credulity, illiteracy, and awe of their adepts to propagate obscurantism, to enact inept or unjust laws, or, as has happened many times in the past and still happens today, to legitimize barbarous practices, preach murderous terrorism, or advocate collective suicide.

We need priests—or better said, *spiritual guides*, so as to avoid the pomp of robes and rites that surrounds the historical image of the priest—to serve as mentors who, without dogmatism or fundamentalism, can inspire, help, and orient. As in any animal collectivity, most members of human societies are entirely occupied with tasks necessary to survival; they have neither the time nor the means to devote themselves to cultural or spiritual activities. Those with the talent and motivation to do so must be encouraged and supported, for they serve an authentic need in the human condition.

Ethical directives are a case in point. Society needs moral rules. But on what foundations are these to be built? For a long time, the laws have been seen as coming from God and promulgated by Him to His people through the voice of prophets. These myths are obsolete. What is left at the most is the abstract notion of good, which, with those of truth, beauty, and love, appear as part of ultimate reality. As to determining what is good and what is bad, the responsibility is ours alone.

According to some, nature provides the criteria for deciding what is good and what is bad. This view is false and can be dangerous. Nature has no moral sense. It knows only the blind law of natural selection, which it applies with utter impartiality to all living species. Our grandeur (if not too grand a word) and our responsibility lie precisely in our power to oppose nature and, if desirable, direct it.

Moral laws are not absolute; they are made by humans to regulate societies; they evolve. Today, in particular, humankind is faced with a host of ethical problems that were unthinkable only a short while ago. These problems can only be solved collectively and consensually, not by authoritarian decisions allegedly ordered by God or imposed by a deified nature.

Recognizing this fact does not invalidate the teachings of the past. It is legitimate today to consider oneself a disciple of, say, Moses, Christ, Confucius, Buddha, or Mohammed, provided this allegiance is not linked with intolerant, proselytizing, and domineering dogmatism. Ethics has its great masters, as do science, literature, art, and philosophy. To follow these masters is in no way demeaning.

Religions should not be abandoned; they should rid themselves of mythical beliefs, irrational pronouncements, obscurantist teachings, magic rituals, claims to superior legitimacy, moral blackmail, not to mention appeals to violence. Cleansed of all these trappings, but with sacredness left, they should be supported and safeguarded, to help us contemplate mystery, respect ethical precepts, celebrate festivities, share joys and sorrows, bear hardships.

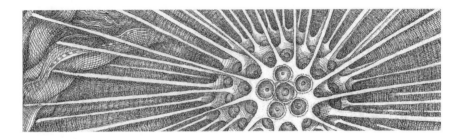

Much has changed, in and around me, since the day when, in front of a burning campfire, I first became aware of the mysteries of the universe. The naïve beliefs of my childhood have been severely shaken. But my sense of wonder remains unaltered. My whole life as a scientist has been permeated with the conviction that I was participating in a meaningful and revealing approach to reality. I have experienced the joy of learning, the almost voluptuous thrill of understanding, the rare flash of illumination, the austere satisfaction of observing the rules of the scientific game, based on intellectual rigor and integrity. I have shared these emotions and imperatives vicariously with other scientists. And I have also vibrated in different registers, in resonance with the poets, writers, artists, and musicians who have moved me by their works and performances. On exceptional occasions, I have felt close to something ineffable, utterly mysterious but real, at least to me, an entity that, for want of a better term, I call Ultimate Reality.

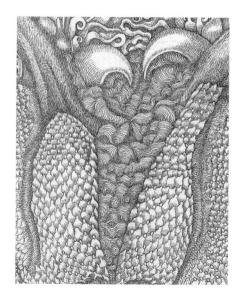

PREFACE

1. One of the founders of experimental physiology and the author of the celebrated *Introduction à l'Étude de la Médecine Expérimentale* (1865), the Frenchman Claude Bernard is often ignored—unjustly, in my opinion—in the Anglo-Saxon world, where his lucid analysis of the experimental approach tends to be eclipsed by the extensive theoretical studies developed one century later by the Austrian-British philosopher Karl Popper, remembered for his felicitous concept of "falsifiability." Added to the well-known cultural divide created by the Channel, Popper's interest in physics, as opposed to Bernard's concern with biology, may be partly responsible for this fact.

2. See my *Vital Dust* (New York: BasicBooks, 1995), pp. 286–91.

3. Ibid., note 2, p. 302.

4. The physicist is Jean Bricmont, co-author of a much-discussed book, by the American physicist Alan Sokal and himself, published first in France under the title *Impostures Intellectuelles* (Paris: Odile Jacob, 1997), and later in the UK (London: Profile Books, 1998), under the title *Intellectual Impostures: Postmodern Philosophers' Abuse of Science*, and in the US (New York: Picador, 1998), under the title *Fashionable Nonsense: Postmodern Intellectuals' Abuse of Science*. Sokal is known for a famous academic hoax in which the editors of a serious American journal, *Social Text*, were tricked into publishing an arrantly nonsensical paper titled *Transgressing the Boundaries: Toward a Transformative Hermeneutics of Quantum Gravity* (described by S. Weinberg, in his *Facing up*, Harvard University Press, 2001, pp. 138–61). Bricmont's article, titled "Science et Religion: L'irréductible antagonisme," appeared in a collection of essays published by the Free University of Brussels, under the title *Où va Dieu?* (Brussels: Editions Complexe, 1999), pp. 247–64.

5. See book in note 2 above.

6. C. de Duve, *Blueprint for a Cell* (Burlington, NC: Neil Patterson Publishers, Carolina Biological Supply Company, 1991).

7. See book in note 2 above, pp. 298–99.

1. Strictly speaking, the kinship demonstrated by the PGK sequence similarities refers to the PGK *genes* of the five organisms considered; it does not necessarily extend to the owners of the genes. This is not just a subtle distinction of purely academic interest. As will be explained in Chapter 7 (see especially note 5), genes do occasionally acquire new owners by mechanisms other than hereditary transmission from one generation to another. What is known of these mechanisms, however, shows that they could not possibly account for the observed PGK sequence similarities. The conclusion that all known living organisms descend from a single ancestral form of life stands firm, supported, not only by the data shown here, but by hundreds of similar cases now on record.

CHAPTER I

1. The last few years have witnessed a certain resurgence of vitalist and finalist doctrines, notably in the framework of the theory of "intelligent design" (see, among others, Chapters 3 and 12). Advocated by a very small minority, this movement is scientifically unimportant; but it is exploited by creationist circles and by a number of other philosophical or religious groups united around the affirmation that "science does not explain everything," that there must be "something else."

2. Turnover experiments can also be performed with substances containing an excess of rare, nonradioactive isotopes, such as carbon of atomic mass 13, hydrogen of atomic mass 2 (heavy hydrogen or deuterium), or oxygen of atomic mass 18. In such cases, the isotopes are measured by means of a mass spectrograph.

3. Originally, the term "oxidation" was invented to designate the *direct* reaction of substances with molecular oxygen (O_2), such as occurs in combustions, carbon giving rise to carbon dioxide (CO_2), hydrogen to water (H_2O), and so on. A more general definition of the term, better applicable to biological phenomena, is *removal of electrons*, that is, of elementary negative charges, and their transfer to some acceptor, which can be oxygen but does not have to be. The simplest electron acceptors are positively charged ions, such as ferric ions (Fe^{+++}), which are thereby converted to ferrous ions (Fe^{++}). When oxygen serves as an electron acceptor, there is a participation of hydrogen ions (H^+), or protons, arising either from the substance that undergoes oxidation or from the dissociation of water molecules. Without entering into details, let it simply be pointed out that the consequence of this complex interplay is that the utilized oxygen ends up in water molecules, whereas water molecules furnish the oxygen in the CO_2 formed. The end result is the same as in ordinary combustions, but the oxygen used travels obligatorily by way of water. This is readily demonstrated with the help of the heavy isotope of oxygen, of atomic mass 18 (see note 2 above).

As can be found in any biochemistry textbook, biological energy transactions rely on a variety of electron transfers between a whole collection of donors and acceptors. In many instances, the electrons travel along chains linking a succession of transmitters that serve as acceptors on one side and as donors on the other, in a kind of electron "bucket brigade." Most often, oxygen serves as the final acceptor at the end of the chains, which explains its central importance. But other electron acceptors provided by the environment, such as ferric iron, may play the same role, for example, for certain bacteria.

In their travels, electrons fall from a higher to a lower energy level, sometimes two or more times in succession, down an "energy staircase." In many such steps, the energy released by the fall of the electrons serves to power a coupled generator of ATP, the common biological energy currency. Because many of these couplings are reversible, electrons can be forced up the energy steps, in reactions called *reductions*, with the help of ATP-supplied energy.

4. The offered description of photosynthetic reductions, like that given earlier for biological oxidations (see note 2, above), applies only to the overall balance of the reactions, not to their mechanisms. In photosynthesis, complex electron transfers account for the reduction of CO_2 and the production of molecular oxygen (the immediate source of which is water).

5. In a greenhouse, incoming light is partly converted into infrared radiation of higher wavelength, which is not allowed out by the glass panes. Carbon dioxide and other atmospheric gases, such as methane, similarly trap solar heat around Earth. Many scientists believe that this so-called greenhouse effect is causing global warming that could be prevented only by decreased use of combustion for energy production, combined with an expansion of the areas planted with carbon dioxide-utilizing greenery.

6. In anaerobic fermentations, the indispensable final electron acceptor (see note 2) is provided by metabolism, so that oxygen—or any other external electron acceptor—is not needed. Thus alcohol or lactic acid is the sugar-derived vehicle by which the electrons released in energy-supplying reactions are discharged into the environment.

7. We shall encounter adenosine later, in the form of its monophosphate (AMP), which is one of the four constituents of RNA and, in a slightly different combination, of DNA, the two major biological information carriers. Adenosine, as will be pointed out, represents a key link between energy and information in living systems. This relationship could be highly significant for the origin of life (see Chapter 4).

8. Occasionally, biological work is not powered by ATP itself, but by an intermediate in the chain of reactions linking the energy-producing metabolic process to the assembly of ATP. The two principal intermediates of this kind are sulfur compounds called *thioesters* and states of electrochemical imbalance across membranes, or *membrane potentials*, the main one of which is created by protons (proton-motive force).

9. Many biosynthetic processes involve reductions, reversing the energy-yielding oxidations.

10. The metabolic reactions that produce energy are mostly electron transfers (see note 2). Many ATP-generating couplings take place by a series of intermediates, one or the other of which may sometimes provide energy for a form of biological work, as mentioned in note 7.

11. The calculation of the protein sequence space is simple, considering that there is a choice among 20 different possibilities for each addition of an amino acid to a string.

Thus, there are already $20 \times 20 = 400$ possible associations of two amino acids, 8,000 of three amino acids, 160,000 of four amino acids; in general 20^n possible strings of n amino acids. This means that 20^{100}, or 10^{130} (one followed by 130 zeros), different protein molecules of 100 amino acids—which are among the shortest in nature—are theoretically possible. It would require a huge number of universes the size of ours to provide enough matter to make just one molecule of each kind. Thus, life uses an infinitesimal fraction of the protein sequence space.

CHAPTER 2

1. The nucleic acid sequence space is calculated exactly like the protein sequence space (see Chapter 1, note 11). For example, with only 300 bases, which corresponds to a protein of 100 amino acids, the number of possible chains (4^n for chains of n bases) already reaches 4^{300}, or 10^{180}, which is a number very much larger than that of the corresponding proteins. This discrepancy is explained by the fact that the same protein may be encoded by many different genes due to the existence of many synonyms in the genetic code.

2. The terminology consisting in designating bases by their initials is ambiguous, because the meaning changes depending on whether information or chemistry is intended. For information purposes, it is customary to use the initials to designate the bases. In the chemical nomenclature, however, these initials designate the combinations of the bases with ribose or, when preceded by d, with deoxyribose. We have already encountered one example of this in Chapter 1, with the A of ATP. This letter does not stand for the base adenine, but rather for adenosine, the combination of this base with ribose. With one inorganic phosphate (P_i) attached to it, adenosine becomes adenosine monophosphate (AMP), which is a nucleotide, one of the four building blocks of RNAs. Adding one or two additional phosphates to AMP, we get ADP and ATP, the main agents of biological energy transfer (ATP is also obtained by the addition of inorganic pyrophosphate [PP_i] to AMP). Similar associations with one, two, or three phosphates exist for all the other combinations of bases with pentoses chemically represented by the initials of the corresponding bases. Thus are known GMP, GDP, and GTP; CMP, CDP, and CTP; and UMP, UDP, and UTP. Similarly, in the DNA line, where deoxyribose replaces ribose, we find dAMP, dADP, and dATP; dGMP, dGDP, and dGTP; dCMP, dCDP, and dCTP; and dTMP, dTDP, and dTTP (it will be remembered that T replaces U in DNA). Note that the other NTPs (N standing for any base combined with a pentose) sometimes substitute for ATP in certain energy transfers. On the other hand, the four NMPs are the nucleotide building blocks of RNAs, whereas the four dNMPs are the building blocks of DNAs. In the biosynthesis of nucleic acids, these NMPs and dNMPs are derived from the corresponding NTPs and dNTPs, the two supernumerary phosphates being released as inorganic pyrophosphate (PP_i). As already mentioned (see Chapter 1, note 7), these facts point to a remarkable and probably revealing relationship between biological energy and information.

3. The term "genetic code" is increasingly misused by the media as a synonym of

"genome," in expressions such as "a person's genetic code." This is an unfortunate source of confusion. The genetic code is the universal dictionary of equivalences between codon triplets in DNA (or RNA) and amino acids in the corresponding proteins. The genome is the sum total of genes present in an individual's DNA. All members of a given species share the same characteristic genome (consider, for example, the human genome, which has been much in the news since its complete sequencing was announced in 2001), but in slightly different versions explainable by the fact that the same gene can exist in more than one molecular form (alleles). If enough such differences exist, as is the case for the human species, each individual is virtually guaranteed to possess a distinct version of the genome of its species. This accounts for the uniqueness of the individual.

4. Not mentioned in our description of the double helix, so as not to introduce unnecessary complications, is the fact that the two strands are antiparallel, that is, have opposite polarities. The head of one faces the tail of the other, and vice versa. They have to be that way for their bases to join.

5. A question that is often raised is whether viruses are living beings. The answer to that question is "no," inasmuch as viruses are incapable of independent life. An unsolved question, however, the object of much debate, is whether viruses descend from primitive intermediates in the origin of life or from complex cells that were, at one time, capable of independent life and have been reduced, in the course of evolution, to the strict minimum needed for reproduction within a parasitized host cell.

6. Contrary to what might be expected, there are not 61 different transfer RNAs, corresponding to the 61 distinct codons. Thanks to possible "wobbling" in the third position of the codon-anticodon joining, certain tRNAs have anticodons that can bind to more than one codon. The number of distinct tRNAs is on the order of 40 (for 20 amino acids).

7. Not all enzymes that join amino acids to their specific transfer RNAs recognize the tRNA's anticodon. Some recognize another part of the tRNA molecule and do not, therefore, strictly speaking, contain one line of the genetic dictionary written into their structures.

8. One advantage of split genes is that they make possible the encoding of several different proteins by the same gene, thanks to *alternative splicing*, the assembly of a gene's exons according to different modalities. It is notably to this process, more than to a particularly large number of genes, that humans owe their wealth of genetic characters. This was one of the great surprises revealed by the complete sequencing of the human genome. It now appears that the human genome contains a bare 30,000 protein-coding genes, only some 10,000 more than that of the lowly nematode worm *Caenorhabditis elegans* and by far fewer than the earlier estimate of about 100,000 and, especially, than the million-odd genes that could be accommodated if all bases were expressed. Beyond the genes, therefore, the science of the future will have to decipher the proteins, whose diversity exceeds by far that of the genes. This new development is expected to lead from *genomics* to *proteomics*, according to contemporary terminology.

CHAPTER 3

1. A school exists that claims, on the strength of sequencing and other results, that the last universal common ancestor (LUCA) was closer to eukaryotes than to prokaryotes, though without displaying all the complex properties of present-day eukaryotic cells. The question hinges largely on molecular kinships among certain genes. This point, which is linked to what is known as the "rooting" problem, will be examined in Chapters 8 and 10. It does not bring back into question the hypothetical portrait of the LUCA sketched in this chapter. In the opinion of the vast majority of investigators, the LUCA was no doubt much closer to bacteria than to eukaryotic cells in its general organization.

2. The existence of early, primitive forms of life is often discussed in the literature in relation with the LUCA. This is a source of confusion that should be avoided. By definition, the last common ancestor must already have possessed all the properties it has bequeathed to its entire descendance. Such an advanced form must obviously have been the outcome of a long history that, starting with highly primitive forms, involved a long succession of progressively more complex forms. In the course of this history, for example, genes and proteins must have gradually acquired the sizes that characterize them today (and already characterized them in the common ancestor). I shall try to retrace this history in the coming chapters.

3. In *The Cradle of Life* (Princeton, NJ: Princeton University Press, 1999), J. W. Schopf has documented in detail the evidence supporting his claim that the traces he has discovered in some ancient Australian rocks believed to be almost 3.5 billion years old are the fossilized remnants of microorganisms related to present-day cyanobacteria, that is, organisms that carry out an advanced form of oxygen-producing photosynthesis. As we shall see in Chapter 8, this claim conflicts to some extent, though not irreconcilably, with the observation that the atmospheric level of oxygen began to rise only some two billion years ago. This question has recently been revived by two articles published side by side in the March 7, 2002, issue of *Nature*. One article, by Schopf's group (vol. 416, pp. 73–76), offers additional evidence reinforcing his claim. The other article, by M. D. Brasier et al., follows immediately (pp. 76–81) under the title "Questioning the evidence for Earth's oldest fossils." In it, a number of data are presented that raise serious doubts on the cyanobacterial origin of the traces and, even, on their biological origin. Discussing the highly technical nature of the controversy is beyond the scope of this book (and of the author's competence). Suffice it to say that there are still uncertainties with respect to the earliest appearance of life on Earth and, especially, with respect to the biological production of oxygen starting more than one billion years before the rise of this gas in the atmosphere.

4. Natural carbon consists mostly of atoms of atomic mass 12, with, in addition, very small amounts of a heavier isotope of atomic mass 13. Thus, the molecules of CO_2 that contain ^{13}C have a molecular mass of 45 (the atomic mass of oxygen is 16), as opposed to 44 for the majority of CO_2 molecules. This slight difference in mass is sufficient to cause the heavier molecules to participate a little more sluggishly in the reaction whereby autotrophic organisms incorporate carbon dioxide into organic compounds. Hence a slight excess of the lighter ^{12}C isotope in the carbon of bio-

logical compounds. So far, no nonbiological reaction has been found to affect the same discrimination. The two carbon isotopes are not radioactive but can be assayed by mass spectrometry (see Chapter 1, note 2).

5. For an explanation of the greenhouse effect, see Chapter 1, note 5.

6. "Cette pâle lueur qui tombe des étoiles," from Corneille's *Le Cid.*

7. Two groups of investigators (M. P. Bernstein et al., *Nature*, vol. 416, pp. 401–403, 2002; and G. M. Munoz Caro et al., *ibid.*, pp. 403–406) have simultaneously published in the same journal the results of almost identical experiments in which mixtures containing water (H_2O), methanol (CH_3OH) and ammonia (NH_3) as main components were subjected to ultraviolet irradiation under conditions of very low temperature (12 to 15 °K) and very high vacuum, close to those that exist in interstellar spaces. In both cases, the investigators have observed the formation of amino acids (three different kinds in the first case; 16 in the second, where the proportion of water was ten times lower). These experiments reinforce the conviction that cosmic chemistry abundantly produces the chemical "seeds" from which life could have originated. Readers will be interested in the commentary by E. L. Shock on this topic (*ibid.*, pp. 380–81).

8. The demonstration, in front of the Académie des Sciences, of the non-occurrence of spontaneous generation (advocated by Félix-Archimède Pouchet) is probably Pasteur's most publicized experiment. A boiled broth kept in an open vessel, but rendered inaccessible to outside air by the narrow, swan-neck shape of the opening, remained sterile, whereas the same broth kept under identical conditions, but not subjected to a preliminary boiling, rapidly developed abundant bacteria. Interestingly, this demonstration of the capital role of sterilization as a means of preventing infection was inspired by Pasteur's vitalist convictions, which caused him to reject spontaneous generation. A hypothesis need not necessarily be correct to be fecund.

9. "Irreducible complexity," as proof of "intelligent design," is the central theme of the book by M. Behe, *Darwin's Black Box* (New York: The Free Press, 1996).

10. For estimates of sequence spaces, see Chapter 1, note 11, and Chapter 2, note 1.

11. W. Dembski, *The Design Inference* (Cambridge: Cambridge University Press, 1998).

12. J. Monod, *Chance and Necessity.* First published in France in 1970, under the title *Le Hasard et la Nécessité* (Paris: Editions du Seuil), and in English translation (by A. Wainhouse) in 1971 (New York: Knopf), p. 145.

CHAPTER 4

1. Even replication of DNA cannot take place without RNA. DNA replication is obligatorily initiated by the synthesis of a short RNA "primer," on which the DNA chain subsequently grows by lengthening. This primer is detached after completion of the newly made DNA molecule.

2. Readers interested in prebiotic RNA synthesis will find a convenient introduction to the present state of this problem in a brief article by L. Orgel, in *Science*, vol. 290, pp. 1306–7 (2000).

3. For the relationship between biological energy and information, see Chapter 1, note 7, and Chapter 2, note 2.

4. The argument that catalytic RNAs could not have supported protometabolism de-

serves some qualification. Several of the major coenzymes involved in both electron and group transfers have a nucleotide-like structure or include a nucleotide in their molecule. To the supporters of the Gilbert version of the RNA world, these coenzymes could be "fossil" remnants of erstwhile ribozymes.

5. Dry heat could have sufficed for peptide formation, as shown by the late American biochemist Sidney Fox, one of the pioneers of origin-of-life research (see Sidney Fox, *The Emergence of Life* (New York: BasicBooks, 1988); see also Chapter 6, note 6). Another possible mechanism is by way of sulfur combinations called thioesters, as I have suggested in my *Blueprint for a Cell* (see Preface, note 6).

6. The multimer theory has been outlined in my *Blueprint for a Cell* (see Preface, note 6). Note that substances resembling my hypothetical multimers are abundantly present in the bacterial world. They are not known to possess catalytic activities, but some of them, for example, gramicidin S and tyrocidin, are antibiotics. The peptides of this family are usually shorter and more heterogeneous in composition than proteins. Interestingly, they often contain amino acids that are not used for the synthesis of proteins, including D-amino acids (see Chapter 5, note 4), and they are made by machineries that do not involve RNAs but use thioesters as intermediates (see note 5, above). In my theory, thioesters are likewise assumed to be the precursors of the multimers.

7. The formation of RNA-like compounds from NTPs is a straightforward reaction that would require only an appropriate catalyst. As mentioned on p. 63, even clays could have done the job. It is also of interest that this kind of reaction is not restricted in nature to the synthesis of true RNA. An enzyme is known, for example, that catalyzes the formation, from ATP molecules, of long chains called poly-A.

8. Consisting of diverse carbon-nitrogen rings, a wide variety of substances, called heterocyclic, could all arise from hydrogen cyanide (HCN) and ammonium cyanide (NH_4CN), which are characteristic products of cosmic chemistry. Thus, adenine—is this why ATP stands out among the NTPs?—can be obtained from ammonium cyanide in a particularly simple reaction, as shown by the Catalonian-American chemist Juan Oro. Traces of this base have also been detected in a meteorite.

9. Natural heterocyclic substances (see note 8, above) include a number of vitamin components, such as nicotinamide, pyridoxal, part of thiamine, flavins, and pterins, that are almost ubiquitously present in living organisms.

10. In a historic experiment, Spiegelman mixed together in a test tube the four precursors of RNA (ATP, GTP, CTP, and UTP), a viral replicating enzyme (extracted from the Qβ virus), and the viral RNA as template. After a number of cycles, in which the same procedure was repeated with the product of the preceding cycle as a source of template, he obtained RNA molecules profoundly different from the starting material. Changing the conditions, by adding an inhibitor of the enzyme, for example, yielded a different kind of RNA as final product. This protocol has been followed, largely unchanged, by subsequent investigators.

CHAPTER 5

1. Amino acids could have reacted with RNAs in the form of some derivative. In particular, amino acid thioesters could have provided the energy needed for the

formation of the RNA-amino acid linkage, as suggested in my *Blueprint for a Cell* (see Preface, note 6).

2. Eigen's claim that the first RNAs may have been ancestral to present-day tRNAs has been referred to in Chapter 4, p. 67.

3. Activation of the amino acids as thioesters (see note 1 above) could have facilitated their attachment to RNA molecules.

4. Some physicists believe the bias in favor of L amino acids to be related to a natural preponderance of this chiral form. Although they have found some weak evidence in support of this theory, a purely physical explanation of biological chirality does not look very likely, as D amino acids, those of opposite chirality, are far from being excluded by life. They are found in many natural substances, together with a number of L amino acids, of which some are not present in proteins (see Chapter 4, note 6).

   A frequently accepted explanation of the chirality problem is that the choice was made by chance. It is assumed that nature played heads-or-tails, so to speak, at the moment the protein-synthesizing machinery was set into place. The coin fell on the L face, but it could just as well have fallen on the D face. Once fate had decided, there was no going back. This could be. But an interesting alternative hypothesis is that the choice was made by the RNAs that initiated protein synthesis, through the simple fact that the geometry of the molecules involved could accommodate type L amino acids, but not type D. Unfortunately, the explanation is not entirely satisfactory, for RNAs are themselves chiral molecules, made with type D ribose molecules. Why not with type L ribose? And, if such had been the case, would D amino acids have been selected? The chirality problem thus remains posed.

5. In present-day life, the site in the transfer RNA molecules to which amino acids are bound corresponds to the end from which the replicating system starts copying RNA molecules. If things were the same in the prebiotic world, the possibility exists that the presence of an amino acid at this end facilitated the interaction of the RNA molecule with the primitive replicating catalyst. It is also possible that this presence rendered the RNA molecule more resistant against degradation.

6. "Evolution in the test tube" has been described in Chapter 4, note 10.

7. As described in Chapter 2, translation depends crucially on the specific joining of amino acids with their so-called cognate transfer RNAs, that is, those containing an appropriate anticodon. Today, this association is carried out by a distinct set of enzymes, each of which possesses a specific binding site for a given amino acid and another for the corresponding transfer RNA or RNAs. In some cases, the RNA-binding site of the enzyme recognizes the anticodon of the transfer RNA but, in others, it recognizes other parts of the RNA molecule. Whether some of the sites now recognized have anything to do with those originally involved in the first RNA-amino acid interactions raises an intriguing question, totally unresolved at the present time.

   Readers interested in delving further into this fascinating subject are referred to the work of the American investigator Michael Yarus (reviewed in *RNA*, vol. 6, pp. 475–84, 2000), one of the few workers interested in RNA-amino acid interactions. Yarus has made the surprising observation that, in certain transfer RNA molecules, the neighborhood of the amino acid-binding site contains a larger number of base triplets corresponding to a codon (not an anticodon!) of the amino acid than would be expected on a purely statistical basis.

8. With respect to the optimization of the genetic code, see the article by G. Vogels in *Science*, vol. 281, pp. 329–31 (1998).

9. The rare deviations from the universal genetic code are found in mitochondria (see Chapter 10), which have very small genomes, usually made of fewer than a dozen genes. These exceptions thus do not contradict the statement, p. 75, that the code can no longer change in a sufficiently complex system. Present attempts at changing the genetic code by means of bioengineering techniques will be mentioned in Chapter 15 (note 13).

10. Only in a system, such as a mitochondrion (see note 9, above), in which the number of genes has been drastically reduced can a change in genetic code exceptionally take place without irremediable havoc.

11. It is probable that the first code was much simpler than the present one, involving only a small number of amino acids and anticodons (as few as four in some models). Code selection would then have taken place in the course of a historical process in which new amino acids and anticodons were progressively recruited. Several models of this kind have been proposed.

12. The code could theoretically be modified by another mechanism besides mutations of amino acid-bearing RNAs. An existing RNA could become adapted to the transport of a different amino acid, for example, through a change in the specificity of a catalyst serving to attach an amino acid to a given carrier RNA (see note 7).

13. The importance of the role played by catalytic RNAs in protometabolism remains moot. As mentioned in Chapter 4, note 4, the existence of a number of nucleotide-containing coenzymes has been used as an argument in support of a richer array of primitive ribozymes. The fact does, however, remain that the synthesis of the first RNAs could not, for obvious reasons, have been catalyzed by RNAs.

14. According to Eigen's theorem, a replicatable molecule cannot exceed in length the inverse of its replication error rate without progressively losing its information content in irreversible fashion through the accumulation of replication errors. Thus, with an error rate of one nucleotide misplaced in 100 (0.01), the maximal admissible length of a replicatable RNA is the inverse of this value, that is, 100 nucleotides. If the molecule is longer, its information will irreversibly degenerate in the course of successive replications. According to this calculation, the length of 75 nucleotides estimated for the first RNAs corresponds to a maximal error rate of 1.33 percent. As a standard of comparison, the best present-day RNA replication enzymes operate with an error rate on the order of three wrongly inserted nucleotides in 100,000 (0.00003), corresponding to a maximal length of about 33,000 bases, which is indeed the order of dimension of the longest viral RNAs. Considering that these enzymes are the products of a very long evolution, honed by natural selection, an error rate of 1.33 percent for primitive replication seems perfectly plausible.

15. The most impressive example of RNA processing is represented by gene splicing, this astonishing process, already mentioned in Chapter 2, whereby some RNAs are

cut into a number of pieces, of which some (exons) are subsequently stitched back together to give the mature molecule, whereas the others (introns) are discarded. Small RNA molecules play a prominent catalytic role in this process.

16. The evoked smothering of a single-stranded RNA by its complementary form is not just an imaginary situation. This phenomenon is exploited therapeutically by the use of so-called nonsense RNA, a form complementary to a functional RNA, which it inhibits by joining with it.

17. Concerning tne minimum number of enzymes required for autonomous life, see paper by C. A. Hutchinson III et al., in *Science*, vol. 286, pp. 2165–69 (1999).

CHAPTER 6

1. The molecules that tend spontaneously to form membranes are called amphiphilic, which, in Greek, means having two loves. The water-loving heads are termed hydrophilic, and the fat-loving tails lipophilic. These adjectives are of general use; they are applied to other molecules, for example, amino acids.

2. Membrane proteins characteristically possess in their chains one or more rod-shaped segments of about 20 largely lipophilic (see note 1) amino acids, which fit snugly within the fatty part of the bilayer, forming what are known as *transmembrane* segments. Depending on the number of such segments, the protein chains run, snake-like, in and out of the membrane, with the two ends of the chain sticking out on the same face if the number of segments is even, and on either side if the number of segments is odd.

3. A number of biological pumps transport electrically charged entities, or *ions*, for example sodium (Na^+), potassium (K^+), calcium (Ca^{++}), chloride (Cl^-), or hydrogen (H^+, protons) ions. Their actions create electric disparities (membrane potentials), which are involved in many crucially important phenomena, including energy transfers, secretion, muscle contraction, nerve conduction, and electric discharges.

4. In relation to the minimum set of enzymes needed for autonomous life, see Chapter 5, note 17.

5. Bacteria with two peripheral membranes are known as Gram-negative because they react negatively to a staining test devised by the Danish bacteriologist Hans Christian Joachim Gram. They are the bacteria that were mentioned on p. 88 as having a wall lined inside by a membrane and thus possessing an external space, the periplasmic space, delimited by two membranes.

6. With respect to primitive cellular envelopes made of protein-like substances, it may be appropriate to recall, for historical reasons, the name of one of the pioneers in origin-of-life research, the American biochemist Sidney Fox, who died in 1999. Fox, already mentioned earlier (see Chapter 4, note 5), made a name for himself in the late 1950s with the discovery that certain mixtures of amino acids exposed to dry heat would polymerize into substances, which he called "proteinoids," that, when mixed with water, formed small vesicles, or "microspheres." He spent the rest of his career studying these structures, which he saw as the original protocells and endowed with a large number of "lifelike" properties, including growth, budding, division, fusion, catalysis, and even motility, excitability, communication, and "sociality." No

one today is ready to follow him to such lengths. But this does not mean that his proteinoids should be dismissed, as they represent amino acid combinations that form under simple conditions. My hypothetical "multimers" (see Chapter 4) could possibly be related to such materials.

7. The light-absorbing pigments, such as chlorophyll, that play the key role in photosynthesis and their connected energy-transducing systems are invariably associated with membranes.

8. The system proposed by Wächtershäuser is supposed to develop on the surface of an iron-sulfur mineral called pyrite, also known as fool's gold because of its appearance, which is taken to be formed in an energy-producing reaction between sulfide and iron. His "iron-sulfur world" has traits in common with my "thioester world," which also rests on the participation of sulfur compounds in the development of life.

9. Wächtershäuser believes, in conformity with the congruence principle, that the early chemistry prefigured present-day biochemistry, without, however, substantiating this belief by the argument presented in the preceding chapter.

10. Bacteria adapted to very hot environments were first isolated from volcanic springs. Deep-sea hydrothermal vents, discovered in the early 1980s, proved a particularly rich source of such organisms, some of which thrive at temperatures as high as 110° C and may even not survive below 80° C. Organisms of this sort have become commercially important as a source of heat-resistant enzymes. The polymerase chain reaction (PCR), now used on a large scale for DNA amplification (see Chapter 15), has been greatly simplified by the use of such an enzyme.

11. A hot, volcanic environment would be a fitting site for Wächtershäuser's "iron-sulfur world" and for my own "thioester world" (see note 8). It also could provide heat-produced phosphate associations, such as pyrophosphates and polyphosphates, believed by many (including a Swedish couple, Herrick and Margaret Baltcheffsky, who have made them the cornerstone of their "pyrophosphate world") to have been precursors of ATP as energy conveyers.

12. The invariable association of photosynthetic systems with membranes has been mentioned in note 7.

CHAPTER 7

1. Sequencing has the distinction of having led to the award of two Nobel prizes in chemistry to the same investigator, England's Frederick Sanger, first, in 1958, for the sequencing of proteins, and later, in 1980, for the sequencing of DNA.

2. DNA amplification, which was first carried out by cloning appropriately programmed bacteria (see Chapter 15), received a tremendous boost in the middle 1980s, with the invention, by the American Kary Mullis, of an almost miraculous, yet extremely simple technique, the polymerase chain reaction (PCR), that allows vanishingly small amounts of DNA to be fished out from complex mixtures and amplified to any desired level. Even traces of DNA extracted from fossil material dating back tens of thousands of years (including Neanderthal man) have been recovered, to be subsequently sequenced, thanks to this remarkable procedure. It has

also become possible, with the help of enzymes extracted from retroviruses (see Chapter 2), to reverse-transcribe rare RNAs into the corresponding DNAs, to be further amplified by PCR.

3. Phylogenetic tree construction has become a minor sideline of a giant, worldwide enterprise. Spurred by the immense potential benefits of the new biotechnologies to food production and human health, not counting investors and patent lawyers, spectacular advances have been made in DNA sequencing techniques. Batteries of automated machines now run out tens of thousands of bases in a single day, feeding them into powerful computers, from which they can be retrieved by means of sophisticated programs that allow the comparison of given sequences with all those that are known, as well as their conversion into the corresponding RNAs and proteins, all in what the French biologist Antoine Danchin has called *in silico*. The complete genomes are already known for a number of bacteria, yeast, one plant, and several animals. The human genome project, which the greatest optimists believed would take at least 15 years when it was launched in 1990, was successfully accomplished just ten years later.

4. Deriving the structure of a protein from that of the corresponding DNA gene is complicated in the case of split genes, in which the expressed parts (exons) are separated by introns (see Chapter 2). In such cases, the analysis is done on DNAs, called complementary, or cDNAs, reverse-transcribed from messenger RNAs, which, as is known, undergo a splicing process that excises the introns and stitches back the exons.

5. As already mentioned in the Introduction (note 1), what is incontrovertibly proven by sequence similarities is the single evolutionary origin of homologous *genes*, not necessarily that of the owners of the genes. It could be imagined that one or more of the organisms studied acquired the PGK gene from an unrelated organism by horizontal gene transfer (see Chapter 3). We shall see later, in this chapter and elsewhere, that this kind of transfer is very common. But the transferred genes come almost exclusively from bacteria (sometimes from viruses), with, most often, bacteria as recipients. Exceptionally, as in the cases of endosymbiotic adoption (see Chapter 10), the genes may be passed from a bacterium to a eukaryotic host that harbors it. It also happens that viruses spread genes among the various organisms they infect. Such mechanisms could, however, in no way account for the presence of the same PGK (in a more or less modified form) in the 19 organisms studied by the investigators, let alone for the hundreds of similar cases now on record.

6. To be precise, the term "point mutation" refers to the replacement of *one base by another* in a DNA sequence. Not all such changes lead to the replacement of one amino acid by another. It will be remembered that the genetic code contains many synonyms (see Chapters 2 and 5). DNA mutations that do not affect the amino acid sequence of the corresponding protein are called *silent*. In reality, the data that are commonly used for phylogenetic analyses are the base sequences of the DNAs. Amino acid sequences were shown in the example for simplicity's sake.

7. The term "genetic drift" refers to an evolutionary mechanism in which random fluctuations lead to permanent genetic modifications without the intervention of

natural selection. According to the Japanese geneticist Motoo Kimura, who has developed a mathematical theory of the phenomenon, this mechanism may actually be more important than Darwinian selection in explaining biological evolution. Few geneticists are willing to follow him that far. But the possibility that mutations that have no incidence on fitness may become stably incorporated into genomes by chance effects is generally accepted. We have seen in Chapter 5 that the genetic code is such as to minimize the deleterious consequences of mutations, which implies maximizing the probability of mutations being either silent (see note 6) or neutral.

CHAPTER 8

1. Woese's name was mentioned in Chapter 6, in connection with the hot-cradle theory, suggested by his finding that all prokaryotes, both archaebacterial and eubacterial, that were identified by comparative sequencing as particularly ancient were thermophiles. We have seen that, even if such were the case—the significance of the sequencing results has been questioned—no inference can be drawn with respect to the environment in which life first originated.

2. Only fairly recently has the name "bacteriologist" been replaced by that of "microbiologist," as the study of microorganisms broadened from its original relationship with infectious diseases to an interest in the organisms themselves, as forms of life in their own right, characterized by a fascinating variety of habitats and metabolic activities.

3. For the importance of horizontal gene transfer, see review by H. Ochman, J. G. Lawrence, and E. A. Groisman, in *Nature*, vol. 405, pp. 299–304 (2000).

4. As seen in Chapter 1 (note 3), the mechanisms involved in oxidation reactions are indirect. The oxygen in CO_2 comes mostly from water molecules, whereas molecular oxygen serves to convert back to water the hydrogen generated in this manner.

5. It has been mentioned in Chapter 1 that heterotrophic organisms generate energy with the help of fermentations, for example, the anaerobic conversion of sugar to alcohol or lactic acid. As to autotrophy, it usually depends on the utilization of light energy or on the oxidation of mineral compounds. But energy-yielding mineral reactions requiring neither light nor oxygen also exist. An important reaction of this kind that could have been of crucial importance in the beginnings of life is the formation of methane (CH_4) from carbon dioxide (CO_2) and hydrogen (H_2), which sustains the vast archaebacterial family of methanogens. Possible sources of the needed hydrogen are known. One that must have been particularly abundant at the time relies on the conversion, catalyzed by ultraviolet light, of ferrous iron (Fe^{++}) to ferric iron (Fe^{+++}). This reaction provides electrons, which combine with protons (H^+ ions) to form molecular hydrogen (see Chapter 1, note 3).

6. In order to understand oxygen-producing photosynthesis—for balance purposes only; this is not a reaction mechanism—take away O_2 from CO_2, adding H_2O to the remaining C, and you obtain free oxygen (O_2) and a CH_2O unit. Substances arising in this manner, whose formula sums up to a certain number of CH_2O units, were, for this reason, called carbohydrates, meaning hydrates (H_2O) of carbon (C). This appellation, although chemically incorrect, is still widely used.

7. Ferrous iron may also have been involved in the generation of hydrogen (see note 5, above). It will be remembered that iron probably played a central role in the origin of life as well (see Chapter 6, note 8). In the present-day world, iron is crucially involved in the utilization of oxygen, as a key component of hemoglobin, the oxygen carrier of the blood, and of the main enzymatic oxidation systems.

8. The main enzymes concerned with the disposal of oxygen-derived toxic substances are superoxide dismutase, which (with the help of hydrogen ions) converts the superoxide ion to hydrogen peroxide; peroxidases, which use hydrogen peroxide to oxidize a number of substances; and catalase, a special peroxidase that possesses the additional ability of decomposing hydrogen peroxide to water and oxygen. Enzymes also exist that repair the damage caused by toxic oxygen derivatives.

9. Oxygen-detoxifying enzymes are described in note 8, above.

10. L. Margulis and D. Sagan, *Micro-Cosmos* (New York: Summit Books, 1986), p. 195.

CHAPTER 9

1. Diatoms are photosynthetic protists, abundantly present in both fresh and salt waters, where they are important components of the phytoplankton. They are characteristically surrounded by a siliceous shell of very fine design. Fossil deposits of these shells (diatomaceous earth) are mined on a large scale for use in the manufacture of paints, detergents, and other commercial preparations, and as filtering aids.

2. In cross-section, the components of the cytomembrane system appear as a two-dimensional network. Hence the name *reticulum* (Latin for small net) originally given to one of its main components (see note 4, below). Three-dimensional reconstruction has shown that the membranes invariably belong to closed sacs, which can, however, like soap bubbles and other similar structures, split into two closed sacs or, alternatively, join together, thus either dividing or sharing contents. Successive fission-fusion events of this kind allow the intracellular *vesicular transport* of materials through different cytomembrane compartments. In the course of this traffic, the materials often undergo various forms of synthetic or degradative *processing*.

3. Import into cells is initiated by *endocytosis*, an internalization process depending on the inward folding, or *invagination*, of the plasma membrane around extracellular objects or substances, followed by pinching off of the invagination into a closed intracellular vesicle, or *endosome*, containing the engulfed material. In most cases, this material is transferred to vesicles, called *lysosomes* (from the Greek *lyein*, to dissolve, and *sôma*, body), within which it is digested into small molecules that are utilized by the cell. Occasionally, the vesicle-enclosed material merely travels through the cell and is discharged outside at a different cell border by a process, termed *exocytosis*, dependent on fusion of the endosome membrane with the plasma membrane in what is essentially a reversal of endocytosis. This kind of transcellular transport, mediated by endocytosis followed by exocytosis, is known as *transcytosis*. It allows bulky materials to traverse tight boundaries made of cell layers, for example, the endothelial lining of blood vessels.

4. Export out of cells begins with the synthesis of the material to be exported, mostly

specialized proteins, in a cytomembrane compartment named *endoplasmic reticulum*, or *ER* (see note 2, above). The proteins are synthesized by ribosomes attached to the outer surface of an ER membrane in a manner such that the growing chain is directly injected into the lumen of the ER sac through a tunnel in the membrane. In cross-section, membranes studded with bound ribosomes have a rough appearance. Hence the distinction between *rough ER* and *smooth ER*, depending on the presence or absence of bound ribosomes. From the rough ER, the synthesized proteins travel through the smooth ER and thence through a characteristic stack of membranous sacs, called the *Golgi* complex, system, or apparatus, often just Golgi (from the name of the Italian neurobiologist Camillo Golgi, who, in the latter part of the nineteenth century, first revealed this structure with the help of special staining techniques). The materials exiting from the Golgi end up being discharged outside the cell by exocytosis (see note 3, above). In the course of their transit from the rough ER to the extracellular milieu, along what is termed the *secretory pathway*, the materials undergo a variety of chemical modifications, including the trimming and refolding of the protein molecules and their fitting with a variety of carbohydrate and other components. The enzymes involved in lysosomal digestion are synthesized and processed in the same way but, instead of being secreted outside the cell (which occasionally happens, physiologically or, more often, pathologically), they are diverted from the Golgi towards the lysosomes, into which they are unloaded.

5. *Actin*, the building block of the most important eukaryotic fibers, is a globular protein molecule fitted at each end with complementary binding sites that allow the molecules to join end to end into threads of indefinite length. Two such threads wind around each other to form a fiber, with special chemical groupings forming the head and tail of the structure.

6. *Tubulin*, which, in two slightly different forms, called α and β, serves to make *microtubules*, consists, like actin, of a globular protein molecule capable of forming single fibers. These, however, thanks to lateral binding sites, assemble further into a tubular structure composed of 13 such fibers in parallel.

7. Intracellular molecular *motors* consist of proteins endowed with the ability to split ATP in a reaction obligatorily linked to a change in their shape, thus forcing the anchoring points of the molecules to move in response to this change in shape. *Myosin* does this in association with actin fibers (see note 9, below), while *dynein* and *kinesin* act similarly in association with microtubules (see note 8, below).

8. The typical 9+2 pattern of *cilia* and *flagella* has a central shaft made of two fused microtubules (see note 6, above) and a surrounding sheath consisting of nine parallel microtubule doublets connected by dynein motors (see note 7, above). Asynchronous contraction and relaxation of the dynein motors cause alternative bendings resulting in beating (cilia) or undulating (flagella) movements of the structure.

9. *Myofibrils* are characterized by a typical parallel arrangement of actin fibers (see note 5, above) and myosin filaments (see note 7, above). In contraction, the myosin filaments move, in ratchet fashion, with respect to the actin fibers by means of a hook-shaped motor appendage, thereby causing the fibrils to shorten. Animal muscles consist of a large number of variously arranged parallel myofibrils.

10. Several severe genetic diseases are due to an inability of peroxisomes to oxidize certain specialized lipids. These accumulate in the cells, notably of the brain, resulting in grave functional defects. One of these diseases, adrenoleukodystrophy, has been the object of a movie, *Lorenzo's Oil*, depicting the struggle of the parents of a young boy afflicted with the disease.

11. The two main peroxisome types characterized by special metabolic properties are the glyoxysomes and the glycosomes. *Glyoxysomes*, which are present in a number of protists, plant cells, and cells of lower animals, owe their name to their role in the *glyoxylate cycle*, a metabolic system that plays a key role in the conversion of fat to carbohydrate. First identified as sites of this process in fatty seedlings, glyoxysomes were later found to belong to the same family as the peroxisomes discovered earlier in animal cells. The glyoxylate cycle system is absent in the peroxisomes of higher animals, which, for this reason are unable to convert fat into carbohydrate.

    Glycosomes are a special kind of peroxisomes so far found only in trypanosomatids, a group of protists that includes the agents of African sleeping sickness, of Chagas disease (a major Latin-American scourge), and of widespread tropical diseases known as leishmaniases. Glycosomes are unique in that they contain the *glycolytic chain*, the central metabolic system that converts sugar to alcohol and carbon dioxide or to lactic acid. This system is situated in the cytosol of all other cell types in which it has been investigated.

12. For details on electron carriers, see Chapter 1, note 3.

13. The exceptions to the rule that proteins are made in the cytosol are the mitochondria and the chloroplasts, which contain ribosomes that accomplish the synthesis of the rare proteins coded for by genes present in these organelles (see Chapter 10).

14. The ribosome-studded membranes of the rough ER are described in note 4, p. 231.

15. For details on the advocates of "something else," see Chapters 3 and 12.

16. On Woese's phylogenetic results, see Chapter 8.

17. For a discussion of the time at which life first appeared on Earth, see Chapter 3, note 3. Even though the microfossil evidence of the existence of living organisms on Earth almost 3.5 billion years ago has been challenged, the great antiquity of life is supported by other fossil evidence and, especially, by the isotopic data mentioned in Chapter 3, p. 45, which point to a date as early as 3.85 billion years ago.

18. Some of the nonmitochondrial catalysts of oxidative reactions, for example, iron-containing proteins called cytochromes, have molecular relatives in the mitochondrial respiratory chain and could conceivably have originated from mitochondria in the first place.

CHAPTER 10

1. Wall-less bacteria, for example the so-called L forms, exist naturally. It is noteworthy that even some bacteria living in very harsh environments may lack an outer wall. Such is the case of the archaebacterium *Thermoplasma acidophilum*, which inhabits waters that are both very hot and very acidic.

2. The volume of a sphere is a function of the cube of the radius, its surface area only of the square of the radius. A growing spherical cell will thus eventually reach a size

above which its surface area will no longer be large enough to allow the necessary exchanges of matter needed for the cell's metabolism. This drawback can be obviated by a change in shape associated with a higher surface-to-volume ratio, or by expansion and folding of the surface. Both changes may have happened to the developing eukaryote.

3. Details on the rough endoplasmic reticulum are provided in Chapter 9, note 4.

4. Recent findings on the ancestry of cytoskeletal proteins are reviewed in a paper by F. van den Ent, L. A. Amos, and J. Löwe, in *Nature*, vol. 413, pp. 39–44 (2001).

5. Remember the "spirochetal nature of intellect," cited in Chapter 8 (see p. 118).

6. For details on the structure of eukaryotic cilia and flagella, see Chapter 9, note 8.

7. The different kinds of peroxisomes are described in Chapter 9, notes 10 and 11.

8. After phagocytosis was discovered, it was found that fluid droplets may be engulfed by a similar process, leading to the coining of the term *pinocytosis* (from the Greek *pinein*, to drink). Still later, the discovery was made that *receptors* on the cell membrane surface are often involved in the selective uptake of extracellular materials, including a variety of proteins and other large molecules. To cover these various internalization processes, which all depend on invaginations of the plasma membrane, the term *endocytosis* was created (see preceding chapter). With the discovery of lysosomes, the endocytic import—often *receptor-mediated*—and lysosomal digestion of extracellular materials came to be identified as a general eukaryotic property. The defense mechanism against pathogenic bacteria discovered by Metchnikoff is a particular instance of this key process.

9. The intracellular association of hydrogenosomes with methanogens is described in a paper by A. Akhmanova et al., in *Nature*, vol. 396, pp. 527–28 (1999).

CHAPTER 11

1. Protection against the consequences of a mutation by the second, nonmutated gene of the pair occurs only if the mutation involved is *recessive*, that is, is not expressed in the presence of a normal homologue. If the mutation is *dominant*, as in the case of Huntington's disease (see Chapter 15), then even a single defective gene suffices to cause a disease.

2. Queen Victoria is the most famous bearer of an X chromosome with the hemophilia defect. Through her children, the disease spread to a number of European princely families.

3. Cichlid fish in Nicaraguan lakes have apparently achieved repeated sympatric speciation, that is, speciation without geographic isolation. See note by M. Kirkpatrick, in *Nature*, vol. 408, pp. 298–99 (2000).

4. In the wake of September 11, 2001, anthrax spores made a dramatic re-entrance on the scene in the United States as potential agents of biological terrorism.

5. The spore that contaminated a bacterial culture in Fleming's laboratory produced a mold that attracted his attention because no bacteria grew around it. Identified as belonging to the genus *Penicillium*, this mold has given its name to penicillin, the first antibiotic. This discovery sparked a worldwide search for other antibiotic-producing molds and mold-like microorganisms, giving rise to a whole family of

mycins (from the Greek *mykês*, mushroom). These include many effective therapeutic agents (streptomycin, aureomycin, gentamycin, kanamycin, neomycin, adriamycin, etc.) and also a number of compounds that turned out to be too toxic for medical use but, together with the drugs, became invaluable aids in research as inhibitors of specific processes, for example, oxidative phosphorylation (antimycin A), DNA transcription (actinomycin D), or protein synthesis (puromycin). Without the discovery of penicillin, not only would many infectious diseases still be the scourges they were in the past, but modern biology could well not yet have made the immense strides that have led to our present understanding of life.

CHAPTER 12

1. "The assumption so freely made by astronomers, physicists, and some biochemists, that once life gets started anywhere, humanoids will eventually and inevitably appear is plainly false" (George Gaylord Simpson, *This View of Life* [New York: Harcourt, Brace & World, 1963], p. 267).

2. "An evolutionist is impressed by the incredible improbability of intelligent life ever to have evolved" (Ernst Mayr, *Toward a New Philosophy of Biology* [Cambridge, MA: Harvard University Press, 1988], p. 69).

3. For a quotation from Gould, see p. 184 in this chapter.

4. The reference to Monod's saying is given in Chapter 3, note 12.

5. M. Denton, *Nature's Destiny. How the Laws of Biology Reveal Purpose in the Universe* (New York: The Free Press, 1998).

6. Ibid., p. 275.

7. Ibid., p. 281.

8. Ibid., p. 281.

9. Ibid., p. 362.

10. K. Miller, *Finding Darwin's God* (New York: HarperCollins, 1999).

11. There are 21 possibilities with two dice if both dice are rolled together. Not all combinations have the same probability. The probability is 1 in 36 for the six doubles, 1 in 18 for the 15 other combinations.

12. The relationship between the probability of occurrence of a given event and the number of opportunities provided for this occurrence is readily calculated. Let P be the probability of the event *not* occurring, then the probability of the event actually taking place is $1 - P^n$, with n being the number of trials.

13. In order to estimate the probability of a point mutation due to a replication error, one assumes that the total number of possible point mutations is $3 \times$ genome (3 different bases may be wrongly inserted at each position). The average number of mistakes (with an error rate of 10^{-9}) is equal to $10^{-9} \times$ genome per cell division. Thus, the probability of a given point mutation is $10^{-9}/3$ per cell division. Hence, the value of P in the equation of note 12 is $(3 \times 10^9 - 1)/(3 \times 10^9)$ per cell division.

14. The fact that a *given* point mutation has a 99.9 percent probability of taking place does not mean that *every* possible point mutation will have occurred in the clone with a 99.9 percent probability. I am indebted to Ivar Giaever for pointing this out

to me and to Jacques Berthet for an in-depth examination of my theory and of Giaever's criticisms.

15. The first medical application of DDT was in the treatment of typhus, which is transmitted by lice. In October 1943, a typhus epidemic in American-occupied Naples was completely mastered in three weeks by treating 1.3 million people with the insecticide.

16. Lucy is the name, inspired by the Beatles song, *Lucy in the Sky with Diamonds*, that was given to the young female australopithecene whose remarkably preserved skeleton was discovered in 1974 by Johanson and Coppens, in the Afar region, in Ethiopia.

17. S. J. Gould, *Wonderful Life* (New York: Norton, 1989), p. 14.

18. S. Conway Morris, *The Crucible of Creation* (Oxford: Oxford University Press, 1998).

19. Ibid., p. 202.

20. Eviatar Nevo, *Mosaic Evolution of Subterranean Mammals* (Oxford: Oxford University Press, 1999). The cited passage is the book's subtitle.

21. On Lucy, see note 16.

22. For details on the time left for life on Earth, see Chapter 16, note 1.

CHAPTER 13

1. One cannot sufficiently insist on the extraordinary slowness of chipped stone evolution: almost 2,000 millennia for the progressive shaping of stones, as opposed to less than ten millennia needed for the change from ox-driven carts to automobiles, and less than one-tenth of a millennium for the transition from an automobile to a lunar spacecraft, or from an abacus to a supercomputer.

2. On whale "culture," see article by M. J. Noad et al., in *Nature*, vol. 408, p. 537 (2000).

3. Sue Savage-Rumbaugh, personal communication.

4. E. O. Wilson, *Consilience, The Unity of Knowledge* (New York: Knopf, 1998).

5. The term "communication" is used here in the sense of "information transfer," whether by active (learning) or passive (imitation) means.

6. R. Dawkins, *The Selfish Gene* (Oxford: Oxford University Press, 1976).

7. Ibid., p. 206.

8. D. Dennett, *Darwin's Dangerous Idea, Evolution and the Meanings of Life* (New York: Simon & Schuster, 1995).

9. S. Blackmore, *The Meme Machine* (Oxford: Oxford University Press, 1999).

10. Ibid., p. 50.

11. Ibid., p. 93.

12. Ibid., p. 99.

CHAPTER 14

1. G. M. Edelman, *Neural Darwinism* (New York: BasicBooks, 1987). See also, by the same author, *Bright Air, Brilliant Fire* (New York: BasicBooks, 1992).

2. J.-P. Changeaux, *Neuronal Man: The Biology of Mind*, trans. L. Garey (New York:

Pantheon, 1985), p. 272. Original edition: *L'Homme Neuronal* (Paris: Fayard, 1983), p. 359.

3. F. Crick, *The Astonishing Hypothesis: The Scientific Search for the Soul* (New York: Scribner's, 1994), p. 3.

4. From book in Chapter 13, note 8, p. 237.

5. Ibid., p. 246.

6. R. Descartes, *Discours de la Méthode* (Paris: Editions de Cluny, 1938), p. 147 (my translation).

7. In rough approximation, the distances of the planets from the Sun vary between 40 times (Mercury) and 4,000 times (Pluto) the Sun's diameter (100 times for Earth). The distances between atomic nuclei in solid matter are on the order of at least 0.2 millionths of a millimeter (2 angstrom units), which is some 50,000 times the average diameter of the more common atomic nuclei.

8. J. Guitton, *Dieu et la Science* (Paris: Grasset, 1991), p. 182 (my translation).

9. Ibid., p. 184 (my translation).

10. For Miller's book, see Chapter 12, note 10.

11. For information on microtubules, see Chapter 9, note 6.

12. Proton-motive force, in particular, which serves as transducer in oxidative phosphorylations, is capable of energizing certain biological phenomena by itself (see Chapter 1, note 8).

13. Cellular motor organelles are described in Chapter 9, notes 8 and 9.

14. About Lucy, see Chapter 12, note 16.

CHAPTER 15

1. As an example, ECoRI, one of the earliest discovered restriction enzymes, cuts DNA at the following site:

$$\begin{array}{l} G|A A^* T\, T\, C \\ C\, T\, T^* A A|G \end{array}$$

Note the palindromic structure of the sequence (the two sequences are identical, read in opposite direction). The asterisk indicates the site of methylation (A) that protects the molecule against splitting (modification).

2. On DNA amplification, see Chapter 7, note 2.

3. On viruses, see Chapter 2.

4. From the Greek *sôma*, body, the term "somatic" designates any cell in the body, with the exception of germ cells.

5. Readers interested in the problems created by the transplantation of somatic nuclei will find a good survey of the question in the article "Nuclear Cloning and Epigenetic Reprogramming of the Genome," by W. M. Rideout III, K. Eggan, and R. Jaenisch, published in *Nature*, vol. 293, pp. 1093–98 (2001).

6. Instead of "organic," the French use the equally quaint term "biologique," "bio" for short, to qualify "natural" foodstuffs.

7. On lysosomes, see Chapter 9, note 3.

8. We have seen in Chapter 11 how diploidy allows many recessive genetic defects to

be harbored without harm (note 1) and how human males may lack this protection for X-linked deficiencies (note 2).

9. Discoverers of diseases—remember Tay and Sachs (see p. 243), or Creutzfeldt and Jakob, mentioned earlier as the discoverers of the human form of mad cow disease (see p. 36)—enjoy the rare privilege of having their names attached to their discoveries.

10. The moving story of the Wexler family is told in detail in the first chapter (*From Curse to Crusade*) of *Genome* by Jerry E. Bishop and Michael Waldholz (New York: Simon & Schuster, 1990). In an interview with Nancy Wexler related by Robert Shapiro in *The Human Blueprint* (New York: St. Martin's Press, 1991), p. 174, she is reported as saying, referring to the possibility of being tested for the defective gene: "There would be greater distress to know that I had it than there would be happiness to know that I was free of it."

11. Lee M. Silver, *Remaking Eden* (New York: Avon Books, 1997).

12. The announcement, by a group of American investigators associated with a private biotechnology company, that they had successfully carried out the first human cloning made headlines in the fall of 2001. In actual fact, the feat was hardly as spectacular as the announcement implied. In a single case, an enucleated, human egg cell, into which a human somatic nucleus had been transplanted, divided up to a six-cell stage and then stopped dividing. Aimed at therapeutic cloning, for the purpose of producing stem cells, this work is said to be continued. Readers interested in the subject may find a first-hand report by the investigators, together with some ethical and legal comments by outsiders, in the January 2002 issue of *Scientific American* (vol. 286, no. 1, pp. 42–49). Further comments have appeared in the February 2002 issue of the same journal (vol. 286, no. 2, pp. 10–11).

13. The artificial expansion of the genetic code is described in an article by I. Hirao et al., published in *Nature Biotechnology*, vol. 20, pp. 177–82 (2002).

14. Explained in more precise terms—for their understanding, a refresher reading of Chapter 2 may be needed—the experiment went as follows. First, the investigators introduced a CTS triplet into an existing DNA gene (C and T are two normal DNA bases; S is a new artificial base, complementary of Y, the second artificial base made by the investigators). Transcription of this DNA in a system containing YTP, in addition to the four normal RNA precursors (ATP, UTP, GTP, and CTP), yielded a messenger RNA with the expected, complementary GAY triplet inserted opposite the DNA CTS triplet (G, it will be remembered, is complementary of C, A of T, and Y of S, as designed). Translation of this RNA was accomplished with a complete system, to which had been added a specially engineered transfer RNA fitted with CUS (complementary of GAY) as anticodon and made to carry an abnormal amino acid (chlorotyrosine). This amino acid was incorporated into the synthesized protein in the position specified by the GAY codon. (It will be remembered that the base pairing with A is T in DNA, but U in RNA. This is why the anticodon in the transfer RNA is CUS, whereas the corresponding DNA triplet is CTS). In summary, a protein containing an unnatural amino acid was synthesized by a biological system engineered so as to handle this amino acid.

CHAPTER 16

1. Five billion years are an ultimate limit for the time Earth is expected to remain fit for life. A number of factors could render it uninhabitable long before the final flare-up of the Sun. One such factor consists in a slow increase of the Sun's brightness, which is expected to double in the next four billion years, progressively raising the temperature of Earth to a level incompatible with the survival of most organisms. Other factors are possible alterations in the tilt, spin, and other properties of our own planet, resulting in drastic climate changes. Upheavals of this sort have occurred in the past and are bound to occur in the future. How soon and to what extent human survival is likely to be affected by such events is impossible to predict with presently available knowledge. It is generally believed that Earth has a minimum of 1.5 billion years left before it becomes severely inhospitable. By then, of course, our successors may have found ways to counteract the anticipated ill effects or to prevent their causes.

2. This is not just science fiction. On January 4, 2000, the British Minister for Science, Lord Sainsbury, announced the setting up of a "Task Force on Potentially Hazardous Near Earth Objects." The report of this Task Force, which addressed various aspects of the problem, including "mitigating possibilities," came out in September 2000 (London: British National Space Center). I am indebted to Sir Crispin Tickell, a member of the Task Force, for a copy of this report.

3. Rereading these lines after 11 September 2001, I cannot help wondering whether my optimism is still justified. Perhaps the prospects of a nuclear holocaust are not as far-fetched as I imagined.

4. The rule that geographic isolation is necessary for the formation of a species (allopatric speciation) does not seem to be absolute. Several cases of sympatric speciation, that is, of the splitting of a given species into two distinct species in the same environment have been described. See Chapter 11, note 3.

5. For reference to Silver's book, see Chapter 15, note 11.

6. J. Glover, *What Sort of People Should There Be?* (London: Penguin, 1984).

7. G. Stock, *Metaman. The Merging of Humans and Machines into a Global Super-organism* (New York: Simon & Schuster, 1993).

CHAPTER 17

1. The past abundance—and even present existence—of water on Mars is almost unanimously accepted. Note, however, that it has recently been suggested, as an "outrageous hypothesis," that liquid CO_2, rather than water, may have carved the valleys of the red planet (see article by Larry O'Hanlon, in *Nature*, vol. 413, pp. 664–66, 2001).

2. Schopf, it will be recalled, has himself come under attack with respect to the authenticity of the microfossils of cyanobacteria he claims to have discovered in Australian rocks almost 3.5 billion years in age (see Chapter 3, note 3). Distinguishing between biological remnants and mineral artifacts in microscopic traces found in very ancient terrains remains a highly problematic assignment.

3. P. Ward and D. Brownlee, *Rare Earth* (New York: Springer Verlag, 2000).

4. Ibid., p. xiv.

5. M. Rees, *Before the Beginning* (Reading, MA: Perseus Books, 1998), p. 38.

6. F. Drake and D. Sobel, *Is Anyone Out There?* (New York: Delacorte Press, 1992), p. 15.

7. M. Gardner, *Science Good, Bad and Bogus* (Buffalo, NY: Prometheus Books, 1981), p. 348.

CHAPTER 18

1. I am referring here to those religions I would, as a scientist, qualify as more enlightened or liberal. Many fundamentalist creeds actually assail scientific knowledge and dispute its validity. It is not the purpose of this book to analyze the many aspects of the conflict that opposes religion and science, especially in the United States. A huge literature exists on the topic.

2. A typical example of such a meeting, in which I participated, is related in *Many Worlds: The New Universe, Extraterrestial Life & the Theological Implications*, ed. by Steven Dick (Philadelphia: Templeton Foundation Press, 2000).

3. Holy Father Message. *Plenary Session on the Origin and Early Evolution of Life (Part I)* (Vatican City: Pontifical Academy of Sciences, 1997), pp. 15–20 (my translation from the French).

4. G. Coyne, Science and Religious Belief, Keynote Paper, *International Symposium on Astrophysical Research and Science Education*, ed. by C. Impey (Notre Dame, IN: University of Notre Dame Press, 1999), pp. 3–10.

5. S. Weinberg, *The First Three Minutes* (New York: BasicBooks, 1977), p. 148.

6. F. Dyson, *Disturbing the Universe* (New York: Harper & Row, 1979), p. 250.

7. Interestingly, the opposition between the two physicists also extends to the domain of religion. Weinberg is known as a militant atheist and unconditional adversary of all forms of religion. Dyson, on the other hand, has written enough kind things about religion to be awarded the lavish 2000 Templeton Prize for Progress in Religion.

8. From book in Chapter 12, note 17.

9. From book in Chapter 3, note 12.

10. From book in note 5 above.

11. From book in Chapter 3, note 12.

12. L. Smolin, *The Life of the Cosmos* (New York: Oxford University Press, 1997).

13. From book in Chapter 17, note 5.

14. S. Weinberg, *Facing up* (Cambridge, MA: Harvard University Press, 2001), pp. 237–38.

15. P. Atkins, *The Creation* (Oxford: Freeman, 1981).

16. R. Dawkins, *The Blind Watchmaker* (New York: Norton, 1986).

17. G. Klein, *The Atheist in the Holy City* (Cambridge, MA: The MIT Press, 1990), p. 203.

INDEX